ANGELO SPADONI

CORPORATE MESSIAH

&

THE FUNDAMENTAL
NATURE OF THE UNIVERSE

WITH INSPIRED KNOWLEDGE THAT
UNLOCKS SECRETS OF THE UNIVERSE
CAN A GROUP OF SEASONED
EXECUTIVES STEER THE WORLD FROM
SELF-DESTRUCTION INTO UNCHARTED
PROSPERITY FOR ALL?

ISBN: 0989000303
ISBN-13: 9780989000307
Library of Congress Control Number: 2013902916
3rd Notch Publishing

Corporate Messiah & The Fundamental Nature of the Universe
Including
The Fundamental Nature of the Universe (FNU) Ongoing Appendix 1.0 (Outline)
www.wikappendix.com

Corporate Messiah & The Fundamental Nature of the Universe

Angelo Spadoni

NOTE: This book includes descriptions of a new branch of intellectual property. If interested in being a part of this effort, including the provision of scientific and financial assistance, please contact the author at www.angelospadoni.com.

This book is dedicated to my family and friends for their enduring love, and to my crew at the shop for continuing to bear with me.

CONTENTS

INTRODUCTION

Thank you for taking the time to read this book.

This book is categorized as a science fiction story; however, the "science" portion of the story is real. It includes a new theory regarding the fundamental nature of the universe that describes how all matter in the universe is moving in a precise spirographic motion, and this motion is the root cause of all known forces and material phenomena. As described in detail, this theory of precise spirographic motion provides a basis for scientific advancement, including a new energy equation that makes $E = mc^2$ obsolete, a mechanistic explanation of gravity, and, a new cymatic model of atomic structure that is superior to the Bohr model.

If you are only interested in the new "science" presented in this book, please read chapters 6 and 7, and the later portion of chapter 8. I acknowledge that this information is not being presented through the normal channels of peer review and publishing for disclosing scientific discoveries. Does it really matter? I can personally attest that this work is the result of over four decades of cognitive effort that includes a formal education in science and engineering from an accredited university.

I took the liberty to include a chapter on the main character's spirituality, because I personally believe it is a more vital topic than any scientific discovery. If you are not interested in reading about what the main character believes and practices in his private life, please skip chapter 4. I am currently setting up a program where rewrites of chapter 4 can be submitted by people of all faiths, including atheism and agnosticism, please visit www.corporatemessiah.com. The selected writers would be paid a royalty, and again, these "rewrites" can be enjoyed with a healthy understanding that each of us has the freedom to practice what we believe in our own home.

If you are mostly interested in the science fiction part of this book, then let me introduce you to Mr. Miles Manta, an honorable man with a unique gift, who is about to pull the trigger launching a meticulously planned effort to save the world from self-destruction.

Sincerely,
Angelo Spadoni
San Diego, California

CHAPTER 1

SUCCESS EXPOSED

The people in the audience were on their feet, and why shouldn't they be? The speaker embodied the most amazing success story imaginable. Everything Miles Manta touched turned to gold.

This was a special audience, for it was on Manta's home turf in San Francisco, California, and the audience was part of a membership organization that he had grown with and been nurtured by. He'd started out very green with this organization, but now he'd become a recognized and distinguished member.

The exclusive organization was called Trek, and it was comprised of success-minded CEOs. The members met at least once per month to discuss and assist each other with common issues faced by CEOs. The premise of the organization was that CEOs are often isolated by their vocation. After all, who better to confide in corporate matters than fellow CEOs? CEOs could not confide in their employees, family, or friends. How could they expect to get an objective point of view on corporate matters from someone who worked under them, especially if it even remotely affected that person's position in the company? They would want family time to be about the family and not work. And, despite the good intentions of friends, pity or envy could affect their relationships. And if friends were like family, it would be inappropriate to involve them with daunting corporate-level issues.

Trek was founded to bring these solitary CEO-types together to receive objective opinions about addressing sensitive issues. Yes, "it's lonely at the top," but the mythical connotation of this cliché is that CEO-types are usually ruthless, overpaid, and self-serving megalomaniacs. In reality, most CEOs are beaten up, overworked, and struggling debt-ridden individuals trying to make payroll and survive in the brutal world of capitalism.

Manta's presentation was an inspirational, down-to-earth talk about taking care of a core market. He suggested companies not overreach their capability, and that real and manageable growth could be achieved by stressing quality over quantity.

The keynote presentation he had just given was not the most important thing on Manta's mind, though. Shortly after his presentation he was meeting with his specific Trek group, Group T9, and it would be his turn to stand before them to describe an issue or problem and receive input and feedback from his fellow highly experienced group members. They had no idea what was coming.

So, after about an hour of hob-knobbing and snacking on heavy hors d'oeuvres outside the lecture hall, the members of Group T9, clad in their polished business attire, slipped into an isolated conference room to conduct their private meeting.

Group T9 consisted of twelve CEOs and the group leader. Among this group was an assorted blend and cross section of cultures, industries, backgrounds, and economic diversity. What they all had in common, though, was that their companies conducted business on a global scale, they were key players in their specific fields of endeavor, and they all had branch offices in the San Francisco Bay Area.

Miles Manta had formed this group about eight years earlier. He had started as a walk-in member of Trek only three years prior. It was unusual, or even unheard of, for a person to rise from new member to group leader status in only three years. But then, no one had put as much time, effort, and ingenuity into improving the operations, scope, and success of the Trek International organization. The board members of Trek recognized Manta's skills and abilities and were excited to see his growth in the organization. Manta had used the best of his persuasive abilities get the Trek corporate brass to allow him to form his own group, and he would have stopped at nothing to get to this moment.

Since Group T9 was formed, the members had all come to know each other very well, and had developed immense trust among themselves. In fact, their mutual trust and loyalty was similar to that of a Roman pact or a Green Beret regiment. It was inconceivable for any of them to betray confidence in any information, whether personal, professional, or about any circumstance or situation.

In addition to group leader Miles Manta, founder of Manta Global, a holding company of various high technology companies conducting business worldwide from their base in San Francisco, California, Group T9 consisted of the following individuals:

1. Bill Oliver: CEO and founder of ListSoft—a computer software company headquartered in Houston, Texas, and San Francisco, California
2. Drew Gardner: CEO of Barramundi, Ltd.—a mining and chemical company based in Australia
3. Claire Wyndham: CEO of the British-based publishing company Wyndham Press, which specialized in news and educational publications
4. Phillip Kruger: a South African global business attorney and head of Kruger Legal International

5. Walter Baroni: the Swiss founder of Baroni Group—a global insurance and real estate holding company
6. Malik Leduc: CEO of Mondeau in France—a global utilities manufacturing and service company
7. Katerina Kovalenko: CEO of KB Energy Systems—an oil and gas service provider in Moscow
8. Carl Zhang: a Chinese industrialist and CEO of Macro Manufacturing, specializing in the manufacture of consumer products
9. Takashi Tanaka: founder and CEO of Intratel—a telephone and Internet service provider in Tokyo
10. Marcos Ramirez: CEO of Omega Energy, an energy services broker from Brazil
11. Rajneesh Patel: CEO of MMC, Ltd.—a diversified manufacturing and trading company headquartered in India dealing in the metal and mineral business
12. Carlos Islas: CEO of Montesa — a Spanish ship building and heavy industry company with facilities throughout Latin America

Each monthly Trek meeting had a different agenda. There might be guest speakers, group work sessions, workshops, or special meetings set aside to address common issues and problems CEOs faced. This particular meeting was a typical group focus meeting where a single group member, or the group leader, got to stand up and describe a problem or issue he or she was facing. Then the whole group would deliberate how best to address the problem or issue.

Today, in Group T9, it was Manta's day to take the floor, and he was eager yet nervous to begin presenting. To conduct the meeting, the chair position was turned over to Bill Oliver, who sat in the chair at the head of the table. With a clearing of his throat and a tapping on the table, Bill called the meeting to order, and all the members pulled their chairs closer and turned their eyes to the front of the room.

"Today we have our distinguished group leader, Mr. Miles Manta, presenting a current issue for our group focus meeting. He gave an inspiring keynote presentation earlier today, and I'm excited to hear what issue he has for us all to help resolve. It is an honor to have him as our group leader and learn from his example." said Bill as he passed around a copy of the meeting's agenda to each member. "Before we start with Miles, however, we have some business to handle. Please take a look at the agenda in front of you."

After about ten minutes of conducting regular group business, including confirming arrangements for the next month's meeting location and speaker, Bill gave the floor to Manta. The man of the day took his position at the front of the room. He stood slightly uneasy before the group. Something was different about him. His breathing seemed irregular. He closed his eyes and mumbled something to himself, as if a prayer, using both hands to brace himself against the table as if to stabilize himself. The rest of the group noticed these odd behaviors.

"Are you OK?" asked Bill. "Do you need to sit down or a drink of water?"

"No, I'll be fine," said Manta as he looked around the table at each member he had groomed for this moment.

With a worn expression, he looked up at the group and said, "I want to disclose to you all that I've been cheating. I've done nothing illegal—that I know of—but I've had an unfair advantage that I've used for my personal benefit. Isn't that the definition of cheating?"

The room was silent. The earnest expressions on the members' faces betrayed the inner feelings that this type of disclosure can cause. Despite long histories of building trusting relationships with others, individuals tend to react differently to this type of confession. Some of the members were thinking about how they should give him the benefit of the doubt. Others saw this kind of admission as a potentially catastrophic situation where one bad apple could spoil the whole bunch, and it needed to be nipped in the bud. Manta knew what was going through their minds.

Bill interrupted the silence. "Miles, how is it that you feel you have cheated?"

"Like I said," Manta replied, "I've had an unfair advantage. I haven't stolen or broken any laws. But…I've been given a gift." Manta paused due to the vulnerability and relief he felt to finally be confessing this to his group. "And I feel I've betrayed the purpose of this gift. I've used it to enrich myself." He paused again looking around the room. "Since I've developed a trusting relationship with all of you," Manta continued, "I would hope you would hear me out and not abandon me, or think I'm insane."

"What in God's name have you done?" asked Bill in a caring but concerned tone.

"OK. Remember taking those mid-term exams in college?" asked Manta. "Well, it's as if I have seen copies of the exams, and I know the questions and answers in advance."

"Are you saying you can see into the future?" asked Drew in his heavy Australian accent.

"Not exactly," replied Manta. "Apparently, I'm *from* the future." Manta straightened his posture and maintained a serious and stern facial expression, as if he was preparing to testify under oath.

THE REACTION

Manta looked around the room and saw many furrowed brows and perplexed looks, and some of the members of the group leaned back in their chairs with their arms crossed looking as if they were sure he was trying to pull off some kind of joke.

"Very funny. Somebody check the calendar. It's not April first, is it?" asked Marcos.

Everyone was now waiting for the punch line. But after an awkwardly long silence, they could see Manta was serious. He wasn't budging.

Now, this was a group of high-level executives where time was money, and in various degrees, all of them were agitated. Some hid their agitation with the expressions of being perplexed, while others were becoming visibly annoyed.

"OK then. Who wants to begin the discussion on how we can help Manta with his problem?" Bill asked, scratching his head and trying to maintain some normalcy.

"This should be easy," said Phillip. "We just ask Miles to demonstrate his ability by telling us who's going to win the Euro Cup match tomorrow."

"Ha, ha! No, I can't tell you who's going to win tomorrow," Manta jabbed back at Phillip, who was known for keeping a close eye on his smart phone for sports updates. "I don't know what will happen tomorrow, just like you couldn't say who would win a gladiator match if you were transported back into first-century Rome. My knowledge is limited," Manta said, trying to keep a patient veneer.

"Well, what can you tell us about the future in general?" asked Walter. "Phil posed a reasonable scenario. You should be able to prove to us that you're from the future with some convincing demonstration we can verify."

"Yes, I can do that," said Manta. "But there are some problems with your seemingly simple request. The main one is that future events may not be fixed. Things could change quite abruptly, so my predictions could be wrong. Revealing information about the future could change events in a serious way. Good things could become bad things. And bad things…well, they could become even worse. Your simple request opens up complex moral and ethical dilemmas that I believe no person or group could resolve. Come on, folks, don't you think I've thought about all this? Don't you think I've considered all of the possible implications?" Manta walked over to the window and put his hands in his pockets as he gathered his thoughts.

After a few seconds, he cleared his throat and came back to the front of the room and continued. "All of you have proven to be game changers in each of your fields of endeavor. If I were to reveal to you specific events, who knows how you would handle the knowledge? In serious matters of life and death, we would all act in ways to preserve our lives and our families. Perhaps your core corruptible human nature might betray your ability to be rational. Yes, you are all corruptible, to some degree. Just as I have been. Remember, we all pledged allegiance to the fundamental belief that we need checks and balances. We've all accepted the fact that we are corruptible."

"Hmmm. That raises another question for me," said Walter. "Why is it that you have this ability or gift you claim to have?"

"Actually, there's the whole gamut of questions of what, where, how, when, and why," interjected Malik, who sat up straighter in his chair and felt frustrated because Manta wasn't offering up explanations fast enough to satisfy his interest.

"Yes, yes, of course," replied Manta. "I've been anticipating this day for years. You know I've become very close to all of you. We've vacationed together. We've gone through various challenges and trials together. Some of us have raised our children together. During all of this time, I've anticipated your reactions to my eventual disclosure.

"I know everything about you and the kind of people you are," said Manta. "We are all familiar with the superheroes found in the comic books. Well, all of you are superheroes; except you are not the make-believe kind. You are the real superheroes this world needs to save itself." Manta looked around the room and into the eyes of his fellow group members who weren't just members, but some of his closest friends. He genuinely cared about these people and he knew they cared for him as well. He had rehearsed how this day would go over and over again in his head and he only hoped he was choosing the best words now that the time had come. He hoped they wouldn't abandon him. He needed their support and connections. He saw confusion in their eyes, but he saw the loyalty as well, which one would see in good friends. He only hoped they would not abandon him now.

Manta continued, "I am concerned that you may feel that I've taken advantage of you. Over the last eight years, I have handpicked every one of you for this purpose. I've spent the last eight years grooming all of you for this day. Why do you think I put you through the exercises of isolating yourselves from your companies with systems in place and protégés ready to step in at any time? Why do you think I had you do the profiling exercises that not only revealed your personality traits, strengths, and weaknesses, but also the traits of those that surround you? When needed, I arranged for infusions of cash to make sure your companies weren't undercapitalized during the downturns. Those consultants and outside investors I introduced to some of you were working for me." Puzzled looks spread across the room.

"Fundamentally you need to know," said Manta, "I have thought through what I believe is the best approach for disclosing the information I'm about to share. None of you should be surprised by this. You know the premeditation required to achieve all I've done in my business enterprises. You know how meticulous and systematic I've been in communicating the needs, goals, and expectations of my business enterprises and the Trek organization. Why would I treat this matter any differently?" Manta asked.

"Yes, we all know how meticulous you are in your business ventures," interjected Katrina, who seemed slightly exasperated. "But we have no idea where you're going with this. I grew up in the Soviet system, and this type of backroom manipulation reminds me of my experiences with the KGB and their level of control, not that I was part of the KGB or anything like that, but it feels as if we were all manipulated in some way," she said in a serious manner.

"I understand how you must feel, and I ask you to indulge me a little longer," Manta responded. "I may be touching on the 'why' part of my disclosure to you today, but I believe there is a reason and a purpose why this has happened to me. It's my understanding that I've been handed a great responsibility. It may sound complicated, but the beauty is that you don't have to understand or even accept my explanation." Manta could hear some members taking in big breaths and letting them out in heavy sighs. He saw some of them exchange glances with each other and he could tell they were trying to forget they felt betrayed a few moments ago.

Manta took in a big breath as well and went on. "It's like a religious belief. We know within our own group some of us adhere to ingrained beliefs of Christianity, Judaism, Islam, and various Eastern faiths, as well as atheism and agnosticism. However, we've never allowed these differences to prevent us and our families from associating and playing together. Why? Because the prevailing basis of our relationship has been the mutual *business* benefit. We have common ground in the basic need of surviving and providing for our families and for the families of our employees."

Manta could see their facial expressions softening. He could have heard a pin drop during his pause as all their eyes were affixed on him and no one stirred in their seats. "I don't need to be from the future to tell you about the importance of modern business relationships. Or how easily business interests can hurdle over racial or cultural conflicts and make friends out of foes, or vice versa. All of you are astutely aware of the significance of today's world of multi-national corporations, instant communication, unhindered transportation, and how the widespread use of English has been used to overcome language barriers. You all know that it's our modern business enterprises that will shape the world as we know it, or destroy it. Again, I'm touching on the why portion of my revelation. I sincerely believe that interjecting critical knowledge to the world of private enterprise may be the main reason I've been sent here to this time and this place."

"Sent by whom?" exclaimed Carl, who spoke with more emotion than normal.

"I will answer that to the best of my ability. But before I do, here's what I would like to recommend, or rather, what I ask from you," Manta said to the group. "First, I trust you'll keep these discussions confidential, which is already stated in the creed of our group. I also ask that I retain possession of the recordings of this and the future meetings that cover the topic we will be discussing. Additionally, I'll do my best to provide a general answer to one specific question from each of you, but I will also retain the option to not answer, or to answer in the best way I see fit. In other words, you can't complain if my answer isn't specific enough. Later you will see why this must be the case. Next, I'm asking that we arrange three eight-hour meetings within the next week to continue this discussion, as there is no way we can get to the heart of what I am trying to convey within the timeframe of this one meeting.

"As you have probably figured out, it was no coincidence that I arranged for our Group T9 family retreat next week at the Napa Estate, and I know that all of your families are already in town. Lastly, I ask that we take a vote at the end of this meeting to verify that we're all in agreement. If any of you are not agreeable to continuing, please disclose this to all of us at the end of the meeting, and you will be excused from further discussions," Manta said.

"This is a major change in direction to the scope and purpose of this group," said Takashi matter-of-factly. "Besides the outrageous and seemingly ridiculous claim that you're from the future, to a large degree, it seems that you're taking our mutually collaborative format and turning it into your own vehicle to push some outrageous personal agenda. I feel like we're being heisted." Nodding heads appeared throughout the group.

"Takashi, I'm not surprised at all that some of you would feel that way," said Manta. "So allow me to put my money where my mouth is. I'm not asking any of you to do anything that jeopardizes your professional or financial well-being. I'm ready to pay whatever you ask. Money is not important to me at this point. I know how important cash flow is to you, and your responsibilities to your company and its employees. I will compensate you all for your time and any lost opportunities. I'm only asking for three meetings within this next week. You have the option to call the whole thing off at any time, and, based on our current bylaws, you can vote me out of this group at any time. You could also vote me into a loony bin if you wish."

"Listen, Miles, we've seen your abilities to persuade the best of them," said Marcos. "I think we should take some time to discuss this among the group now, before we take a vote at the end of the meeting. And, to be honest, based on the unique nature of your presentation, I think the discussion shouldn't include you." Marcos voiced the opinion of the whole group.

"I understand," said Manta. "Remember, though, Marcos, this is my day to communicate my needs to the group. I still have the usual four hours to present my topic today and receive feedback."

"Yes, you're free to speak and ask what you want from us," said Bill. "After all, you did bring the coffee and donuts. I remember back in seventy-six, in my prior Trek group, when a young man presented his reptile collection to the group. I never thought that one could be beat." Bill chuckled as he remembered. "OK, let's see…" Bill continued. "This is all very unusual. Um…OK. I tell you what. Miles, could you please leave the room and give us about twenty minutes?"

"Sure, no problem," said Manta. "I'll go take a walk down the hall and be back in twenty minutes. Oh, there is something else. At about five o'clock this afternoon, there'll be a horrific tragedy involving the use of a nuclear bomb against civilians. I want to assure you that you and your families are safe, and that there is absolutely nothing you or I can do to prevent this tragedy from occurring." With that, he left the room.

CHAPTER 2

TEST AND ACCEPTANCE

Twenty minutes later, Miles Manta entered the room.

"Please take the floor, Miles," said Bill.

"So, what did you decide?" asked Manta as he took his place in the front of the room.

"Well, Miles, you've put us in an awkward situation," said Bill. "Technically, you're correct: you have the floor for four hours—well, now more like three hours—and you're free to speak about whatever you want."

"I will yield my individual time if Miles needs it," interrupted Rajneesh willingly. Rajneesh was a long-time Trek veteran and an authority on bylaws. The group consulted him whenever matters of procedure needed clarification.

"Duly noted," said Bill. "We're obligated to hear your presentation and help you with your issue in the most sincere and frank manner that we're able. We're also obligated to consider your terms. Regarding the comment you made when you left the room about a nuclear bomb…we are all sickened that you would toy with us like that. We can only assume you were trying to get our attention and were kidding, because the reality of such a thing is unfathomable. We have no choice but to try and act as if we did not hear it."

"I understand your reactions," said Manta. "I hope to help you understand my situation, and how my anxiety swells every night before I go to bed. After I explain my experiences, you may appreciate that I had two options. I could either use drugs to force myself to sleep, or I could work myself so hard that I would crash face-first into my pillow each night. I choose the latter. I have no complaints. As you are aware, other than being a workaholic, I have lived comfortably."

The group could understand that, since many worked as hard as he did. What they couldn't associate with, though, were the feelings of anxiety Manta felt. They were eager to know more. Many of the members adjusted themselves in their chairs and clasped their hands together with their elbows resting on their armrests. They were ready for Manta to further explain himself and gave him their full and undivided attention.

"With that," said Manta, "let's discuss the who, what, where, how, when, and why."

Manta stood up straight and tall as he began his explanation. "Let's start with the who, or *me*. Who I really am," said Manta. "As you know, I grew up in a highly accelerated manner. When I was fourteen years old, my best friend was a twenty-eight-year-old college graduate. I remember going to college parties and hanging out drinking beer while having high-level conversations. I knew I was freaking them out.

"I could get away with associating with older people because I was the youngest of ten children. My close friends were my older siblings' friends. Do you think it's strange that, after the age of ten, I never had a friend that was even close to my age? I could not relate to their ignorance of the world around them, and they certainly could not relate to me. Also after having ten kids, my mother practically did not even know I existed, where I was, or what I was doing. She always assumed my older brothers and sisters were watching me."

Manta paused for a moment as he thought about his boyhood, his freedom, and his ability to be so invisible and innocently non-threatening to those around him. He continued, "By the time I reached my early twenties, I was like a square peg trying to fit into a round hole. I could not and did not have any interest in developing relationships with people. When I slept, I began to have dreams about places and people that I knew and that knew me, but they were not part of my real waking life. Then I started to dream during the day, while I was awake. But I wasn't really dreaming; it was more like inspirations. I was thinking clearly about people I had never met and places I had never visited. I would have frequent déjà vu experiences that would nearly disable me. I tried to relate these experiences to others, including professional psychiatrists, but no one could help me or explain what was happening to me.

"I say this humbly: one very powerful feeling that I had was an amazing sense of self-confidence," said Manta. "I truly felt that I could do anything that I wanted to. All I had to do was put my mind to it and get to work. Sounds pretty simple, doesn't it? But it was different for me, mostly because whatever I took on, I found I knew more than most of the reference materials I studied. My automatic mastery of various subjects alienated me from a lot of my peers and professors, and I came off very threatening to many high-level people. Never outshine the master, right? Naturally, people expect you to have tenure before you claim to be an expert.

"I don't want you to think I am an angel, but this self-confidence was tied closely to a sense of morality. I inherently knew that if I betrayed the purpose of this gift all would be lost. I hope you understand why I have so carefully planned and anticipated my disclosure to you. Nearly everything that I have done in the last eleven years has been to get to this day—to make this revelation to you. To free myself from the feeling of guilt that I have not lived up to my purpose.

"I had one dream in particular that was especially real and vivid. I dreamt I was sitting in a chair in a room with a big television screen. My body was secured to the chair and I had no sensation of my arms or legs. A woman entered the room and was standing in front of me, between me and the large television screen. She was trying to talk to me. Have you ever had the kind of dream where you were trying to run, or throw a punch, and your limbs would not work?" A few of the group members nodded their heads and mumbled "yeah."

Manta continued. "Well, in this case, it was not just my limbs. I had no ability to speak. I was trying to communicate my needs to this person. I could understand every word being said to me, and I knew that this person was making every effort to understand me and help me. I was straining with all of my ability, though, trying to tell this person that I cared deeply, and describe how much I loved this person, and the others that would regularly visit and render care for me. But it was an exercise of futility. I could only grunt and shake my head violently. It was obvious to me that the other person was as frustrated as I was. The dream troubled me for years.

"I had a defining moment when I was twenty-four years old. It was a Saturday morning, and I was sitting in my living room feeling a general sense of melancholy, as I often did, when my doorbell rang. I answered the door, and a religious person was standing there offering me Bible literature. He said that their publication gave prophetic descriptions of events that would occur in the near future. He asked me what I thought the future held for humankind.

"At this point, my competitive spirit came out and I confronted him with my own set of questions, including questions about who authored the Bible. In response, he opened up what he called an inter-linear translation of the Bible, which included a word-for-word English translation of the Greek text. The line of English was written just below each line of Greek. To impress me that his opinion was bible-based he began to answer my question by reading a verse of the Greek text in what was apparently his first language.

"This was the first time that I had ever seen written Greek. And I had never in my life heard the Greek language spoken out loud. Greek is simply not heard very often in the United States.

"But I realized, listening to him and looking at the Greek text, that I could understand it fluently.

"Now, understand this: even though I could master many subjects, language arts was not one of them. I had only taken two years of Spanish in high school. I did well in the classroom, and on the street I could get by, but I was by no means fluent in Spanish.

"Imagine my shock when I spontaneously spoke Greek to this person. It was as if all the bottled-up frustration that I had experienced in that insanely vivid dream was immediately uncorked. I was consumed with deep, gut-wrenching fear. I felt extremely nauseated. The last thing I remember was reaching for the railing on my front porch.

"When I came to, I was on a gurney and being tended to by an EMT. Apparently I had collapsed and smacked my face as I rolled halfway down a stairway off my front porch. There was blood all over. I'll spare you all the details of my injuries.

"You hear about people having religious transformations; or maybe you have had one yourself," said Manta. "You hear about the celebrities like Cassius Clay becoming Muhammad Ali and the Beatles adopting Maharishi Yogi, and of course there are millions of lesser-known individuals suddenly making radical conversions. Well, I do not want to scare you all away, and I promise not to try to convert you, but relating my experience to you as a spiritual awakening is the best example I can use to describe what happened to me. I want to make it clear that I do not want to have any religious discussions with any of you; I fear that I could alienate you by interjecting religious bias, and I insist we keep our religious beliefs separate from the group.

"My apparent religious transformation was obviously more radical than most," said Manta. "Over the course of many months, I had visions and dreams. There was a tugging and a pulling. In my dreams I heard heavy breathing like an enraged bull. I was taunted by visions. I saw blood-red skies—the deepest color red you could ever imagine. It was an extremely frightening time for me, but in the end I came to my own conclusion about who I was and from where I had come.

"I spent a considerable amount of time and resources in Greece and Macedonia searching for answers. I have become quite the expert on Greece and Greek culture. Or maybe I was an expert before I started. Either way, from my literature search or my travels, I have never found anything in a material sense that has shed light on my history."

"What do you mean 'in a material sense'?" asked Carlos skeptically.

"I mean physical things—like cities, houses, plazas—that I could identify or knew that I had been to before. Even the Parthenon doesn't seem intimately familiar to me or as if I identified it with my nationality," said Manta.

"So if you really want to know who I am, I believe I am…or was…a person afflicted with a severe case of an autism-like condition who lived in a Greek-speaking country about four hundred years in the future. My knowledge of history is mainly from sitting in front of that television screen for practically my whole preexistence. I was probably placed in an institution and never traveled beyond the outer walls. Who knows…" said Manta.

"Who I was or where I came from are not important; what I believe is the reason for me being here is that I apparently have come as a messenger," revealed Manta slowly. He anticipated that this part of his disclosure would be treading on thin ice.

"Miles, I must say that you don't really seem certain about what you are saying. For example, you say 'apparently' that is why you have come. That does not sound very convincing," interrupted Drew.

"Good point," said Manta. "I said 'apparently' because I really don't know. I am not sure about a lot of things. Sometimes I am sure about who I am, then the next minute I feel as though I am flying by the seat of my pants. Sometimes it seems as if I only understand enough to help me get through the next moment. Other times I am so focused that I really can see things before they happen. Like I said, my experience is somewhere between a dream-like state and a state of being intensely inspired.

"You have to understand," said Manta, "what I am telling you are my own interpretations. Maybe I did not come from the future. Maybe somehow this whole idea was planted in my head. I can only describe it to you the way I have experienced it, the way I have interpreted it. Yet, in the end, it doesn't matter who I am or where I came from. What matters is the importance of what I have to tell you. If you have a better understanding of why I am here, or what is causing me to have these experiences, please let me know.

"So, with that being said, let me explain to you my interpretation of why I am here, and what I believe is the right thing to do," said Manta as he walked to the side of the room, picked up the lectern that was against the wall, and placed it in the front of the room to put some of his notes on.

"First, something has gone seriously wrong with the world today. Whether you believe in Bible stories or not, let's use them as examples because they are well known and widely used. Remember the story about when the people of the world became so wicked that God caused a flood? Or do you remember the story about a group of people enslaved by a mean Pharaoh, and the God of the Bible caused things to happen, like fire coming out of the sky and the sea splitting in two, to allow them to escape? Well, in each case these things *did not* have to happen, except that something went wrong that created a need for direct intervention. Apparently there are times when something has to be done to correct the course of history. I would like to believe these corrections are for the better. I expect a lot of people would disagree with this, especially if your loved ones were drowned in a flood or collapsing walls of water. But, in each case, everyone should have known, or should have had some interest in knowing, about the coming options. For example, so the story goes, everyone knew Noah was building an ark and had an opportunity to join him and build more arks. There were ten plagues preceding the eventual exodus out of Egypt, and apparently a vast mixed company of Egyptians did join the Israelites.

"We simply do not know all of the facts regarding these past calamities," said Manta. "But, from my research and understanding, there was a justification for God's actions according to the laws of the land. There was a legal right for God to intervene and change the course of history if God deemed it necessary. That is why I believe I have been sent. There is a time-line issue that requires me to disseminate the information that I have, so a change can be made. If a correction is not made, the time allowance for human activity on Earth could be altered is such a way as to create a dilemma for a very important person. In other words, he, she, it, they, whatever, or whomever you want to call God, will not allow their schedule to be altered by human activity that prematurely damages Earth.

"Another example from the Bible is the Tower of Babel," said Manta. "In this case, the people knew that they were supposed to spread out and cover Earth. They were defiant of this order and insisted on staying together in one place to build a great tower.

"We can glean a lot of understanding from what the God of the Bible said at that time. He said 'nothing they plan to do will be impossible for them.'

"According to the story, their language was confused so they could not finish the tower, and their plans were thwarted. In other words, something had to be done to slow these people down. If not, humankind would have progressed so quickly that the Greeks or Etruscans could have developed and destroyed Earth with nuclear weapons by the second millennia BCE.

"I believe this incident with the Tower of Babel is closely tied with what has happened to me," said Manta. "Apparently there needs to be another adjustment, and this time I am the messenger. In the case of the Tower of Babel, something was planted in the heads of these people, namely an entirely new vocabulary. In my case, it is information about the laws of science that has been planted into my head. It is possible that Isaac Newton or others were given a similar revelation."

Walter looked around the room at the different members of the group trying to Figureout if he should say something or not. Then he spurted out, "OK, OK, this is all getting too weird. If we are going to continue with this, we need solid proof. You said that we could ask you questions about the future, implying that we could prove the validity of your prophetic knowledge, so can we begin with the questions?"

"Well, not exactly," countered Manta. "You can play back the recording to hear exactly what I said." he realized that he may have sounded slightly aggressive, so he paused and calmly tried to explain, "I may be limited in what I can say for moral or ethical reasons. Or maybe I just flat out don't know the answer." Manta paused to gather his thoughts after the interruption. He wasn't exactly done with where he was going and what he was just saying, but he could tell the group wanted him to move on faster to the question-and-answer portion that he had promised.

"Yeah, OK, let's go for it," said Manta. "Each of you gets one question. Let's start to my left with Malik and then whoever has a question after that, please speak up.

"OK, Malik, you go first."

"Alright," said Malik as he quickly gathered his thoughts on the spot. "Tell us... will there be peace in the Middle East?"

Malik sat back entirely intrigued as to how Manta was going to respond. All the others were equally attentive. It was such a strange feeling for everyone in the group, to start the round of questions and possibly learn the future. They felt as if they were characters in a science fiction movie.

"Yes," said Manta. "There will be peace in the Middle East. What precipitated this peace is a gesture made by a group of people. In fact, in the future, it is universally known as 'The Greatest Gesture' of all human history. For now, I would like to leave it at that. That is all I can risk telling you, and I have to remind all of you to keep that, and everything else I say, in confidence. Can you imagine how terrible it would be for me to provide details and possibly screw up something that is considered one of the most positive turning points in human history? As we go through more questions I will try to come back to this subject with more information that may shed more light on Malik's question."

He felt assured the group understood this from the slow nods he saw around the table. They were quiet and he could tell they were trying to digest the information and grasp the reality of the situation of this meeting. There was a rather long pause after Manta finished answering Malik's question.

Phillip broke the silence by tapping his pointer finger on the table, then raising his hand up to get Manta's attention. "OK, Phillip, it's your turn," said Manta.

Phillip raised his voice so all could clearly hear him from one end of the table to the other. "Miles, I'd like to know if, in the future, criminals are sent to jail. Is there capital punishment? What happens? Because I know that more than one in thirty-two persons are in jail or correctional supervision in the US, and it is projected to get worse. What's going to happen? I mean, is this problem going to sink us?"

"That's a good question, Phil, and I am not surprised you ask, considering your background in the judicial system," responded Manta. "Things will be very different in the future regarding prisons. In the future, the world becomes much more business-oriented, and this relates to a trend that you see now where a lot of government services are contracted out to private companies. In the future, the difference is that these private contractors are bidding on a worldwide basis. There are no restrictions on receiving bids from anywhere in the world. As it turns out, the most successful prison companies are in North Africa. Most of the prisoners from the US are sent to Somalia and Ethiopia. Prisoners are chipped for identification purposes, and they become subjects of the respective country to which they are sent. So, as you can imagine, in the future, going to prison is something to fear, and prisoners are treated like a commodity. But it has become a great and flourishing business with a healthy amount of competition and trade. For some reason, the Northern Provinces of the African continent seem to conduct the business better than anyone else. When you think of these areas today, it's hard to imagine with all the corruption, piracy, and poverty, that doing legitimate trade is possible. In the future it is very different, and the competitive market for prisoners has resulted in the highest quality facilities and programs, unmatched anywhere in the world. In the future, court systems have no problems sending people to prison because they have so many options to choose from. The court systems continuously review and inspect the facilities and can recall prisoners, or ship them to another competing institution, at any time. Incentives are designed into the system for prisoners to want to get out of prison and back to their home country. Also, there is a great incentive for would-be criminals to obey the laws so as not to go to prison in the first place."

Manta's expression changed and he took a drink of water and looked out the window. The group could tell that was the extent to Manta's answer and he was ready to move on to the next question. Claire, who had been quiet thus far during Manta's presentation, spoke up from the middle right hand of the table. "Miles, something that I happen to be interested in is education. Are there still kindergarten-through-twelfth-grade schools and colleges and universities? Or do we figure out how to just inject knowledge into people's heads or take a pill or something?"

"Claire, my friend, that is a great question. And I am not surprised you ask because I know your passion and career have been closely tied to improving education, and you know that it is constantly changing," said Manta. "Education is incredibly different in the future. In fact, I would say that of all the real changes between now and my time in the future, education is one of the most radically different, especially in pre-school through sixth grade.

"There will be major changes in the world, which will include worldwide revolutions. Historians will name one the 'Garden Revolution.' You probably can surmise that it is sort of the opposite of the Industrial Revolution. Now I was not a historian in the future, so what I tell you now is what I learned and not necessarily a complete historical account.

"It should be obvious to you that the concept of the industrial world cannot continue as we know it. Really, the mass consumerism we know today stemmed from the two World Wars. Yes, Henry Ford built the first assembly line and paid workers enough to buy the products they were making, and this started us on the road to the consumerism, but it was the World Wars that put the real mess in motion. When you build factories to produce millions of warplanes, tanks, and guns, what happens when the war ends? Well, you keep building and create consumer-based economies, which eventually consume all of the world's resources at breakneck speed. Hence, eventually there will be a crashing halt to the industrial-based way of life. The Garden Revolution resulted in the downfall of suburbia, after it was realized that the suburbs were designed to make people dependent on cars and wasteful services. People flocked back to either urban life or they moved to farm communities. Sprawling suburbs were reclaimed as farming areas, community gardens, or restored to natural habitats, dispersed with small, high-density urban areas similar to the hill towns of Tuscany. But that is another story.

"To answer your question about education, though: in a nutshell, the Industrial Revolution, and all that went with it, created a lot of dysfunctional and unhappy children by diminishing the importance of the basic family unit. Low and behold, the survival of the human race depends on stable and functional children. One of the main legacies of the Garden Revolution is the standard practice that practically all children from the ages of three to twelve years old are raised in farm-like communities that harvest crops and raise animals. In the case where the family's main residence is located in the city, the city parents relocate to the farm to live with their children during these formative years. The importance of these formative years includes forming healthy eating habits and, surprisingly, more contact with germs. The soil and animal exposure in a farm environment helps the children form healthy antibodies, resulting in tremendous reductions in health care costs for society. Yes, it is taken that seriously. After this primary education, children have a wide variety of choices for middle school and high school. Many continue in farm-related work, many continue in trade schools, and only some pursue academia. The pursuit of academia is definitely not as highly touted in the future as it is today. Going to medical school in the future is considered a trade school. The basis of grounding the children with an upbringing in agriculture is that it gives all people a common connection to the natural processes that sustain the Earth. In the future there is no excuse for unemployment or homelessness when we all know we can work, eat, and have shelter on a farm. And going back to the farm, if you have hit hard times, is not perceived as shameful or lower class. Many gainfully employed people make a regular practice of 'going back to the farm' for the health and spiritual benefits it can provide.

Manta finished answering the question, stood next to the lectern, and leaned his arm on the side of it. Katerina spoke up next from the other side of the table, very straightforward. "Miles, will the world run out of oil? And what will be the makeup of our energy consumption?"

"No, we do not run out of oil," began Manta. "The price of oil constantly fluctuates depending on supply and demand. Most important, there is an abundant energy supply available from a variety of sources, and we will talk more about that later. We become completely weaned from our dependence on crude oil. As far as liquid oil products go, those derived from biomass are popular because they can be more easily transported and stored. Instead of refineries, there are 'digesters' that convert biomass to liquid oil products, including fuel. The digesters are relatively inexpensive to build, so they can be installed where the biomass supplies are located. Electric vehicles are popular in some communities in the future, including personal vehicles called 'tram cars' that run on a network of elevated rails and derive their power from the rail itself, similar to today's trolley cars. The elevated tram car rails are nice because they do not interrupt the migration paths of the animals, and during the wee hours of the night automated delivery vehicles transport products to and from the urban areas, so there is much less congestion during the day. Surprisingly, personal cars that operate on liquid bio-fuels are popular with enthusiasts because of their performance, independence, and nostalgia. Understand that the low-pressure combustion systems of the future are much cleaner than those used today, and the average liquid fuel commuter vehicle can get about a hundred miles per gallon. Also, the whole automobile business in the future is different because not as many people own cars, and the infrastructure of roads is not what we see today because they are simply too costly to maintain. Urban lifestyles and public transportation systems are so superior that owning a car in the future becomes like someone owning a horse today. Yes, there are car enthusiasts, but the cars are more expensive and messy, compared to the alternatives."

Manta changed positions again and stood behind the lectern with his hands in his suit pockets. Rajneesh was the next to voice his question. He cleared his throat and spoke up from near the end of the table.

"Miles, what is the future of religion? I know you said you had some kind of religious experience. But there is a widespread point-of-view in this country that religion is becoming obsolete and is a root cause of wars and division. What happens?"

"Great question, Rajneesh," replied Manta. "Religion definitely does not end. In fact, it becomes as big and organized as ever. There is much more tolerance between people of different beliefs. In the future, to judge someone strictly on religious faith would be similar to showing up at a public event today clad in swastikas. It will be unheard of.

"Visiting people and places of different cultures and religions is one of the most frequent pastimes. One of the major changes in the future is something you might find to be somewhat comical. In the future, religion merges with tourism." Once Manta said this, he saw many heads tilt and members stifle laughs under their breath. He knew he must be sounding somewhat crazy to them, but he was surprised how well they were taking in all of this new information.

He continued. "What happens in the future is that most of the well-established religions form what are called 'temple cities.' You could guess where some temple cities are located: the Mormons have Salt Lake City, the Catholics have Rome, the Jews have New Jerusalem, the Hindus have Puri in India, and the Muslims have New Mecca. These cities mostly cater to holiday travelers and tourist organizations. Any person who visits these cities is treated with the greatest care and consideration. These cities are the principle means of converting people to a religion. In fact, there is generally no proselytizing outside of temple cities. How much people enjoy their vacations is a reflection of the host religious organization itself. And they are all very good at serving the tourists. The temple city model is really in sync with a fundamental need most people have for displaying spirituality, but it is all on a level playing field related to a purely secular and cultural experience.

"When you visit a temple city, you really feel like you are visiting people in their personal homes. Even today, we know that it is a common trait for people to enjoy visitors into their homes, and to make sure those visitors are well taken care of. This trait is especially inherent with the people that live and work in temple cities. This has proven to work very well to preserve the customs and core values of humanity's different religions and lifestyles, and to share these cultural roots with others. An interesting characteristic of temple cities is that the best real estate is dedicated to accommodating visitors. For example, you never see private homes on the waterfront or in the prime view areas in temple cities. These prime real estate locations are reserved for visitors and temple city residents that enjoy vacationing in their own temple city. This does not mean that the prime real estate areas are necessarily developed with large hotels. In many cases they are preserved as natural habitats, with limited traditional dwellings, that demonstrate how people can coexist with the natural flora and fauna.

"There are people who immerse themselves in the temple city culture. Some are called 'hoppers,' because they will live as devout Catholic one year, then the next year they will live as a Hindu monk. This is not done with a true religious conversion; it is purely a cultural experience. Just like a musician who switches from Indonesian gamelan music to Mexican mariachi music, the whole wardrobe and antics change. It's about different cultures, and what makes people appreciate this is recognition that we all have a common origin. That origin is our common roots to the beginning of civilization in what today's scholars call Mesopotamia.

"In the future, the common threads between different faiths become more apparent. The simplest example is how Jewish and Muslim faiths both identify Abraham as their patriarch, where the Muslim people descended from Abraham's first born son Ishmael, and the Jewish people descended from his second son Isaac.

"And, of course, religious tourism in temple cities is a profitable business that keeps the wheels of commerce and trade moving. Because people's livelihoods are at stake, there is an inherent safety factor with this arrangement as well. For example, in some of the more traditional temple cities, in what are now perceived as places a Westerner would never visit, if a person were caught threatening or stealing from a tourist, a local tribunal could have the person stoned to death. It rarely happens, and there are checks and balances, but it can happen. The point is the residents of a temple city take the safety of their visitors very seriously. I don't mean to make it sound so rigid. Just think of it as Disneyland, where there is a lot of focus on free flowing family entertainment, but you know behind the scenes security is tight. "

"OK, my turn," said Bill Oliver. "Does the world get destroyed by global warming? Do the ice caps melt?"

"No, the world does not get destroyed by global warming," said Manta. "I think the answer to your question is actually one of the main reasons why I was sent."

Manta paused and looked up at the clock on the back wall. It was 4:50 p.m. "Time to turn on the TV. Please turn the channel to CNN, Bill."

Bill followed Manta's directions and got up and turned the TV on in the front corner of the room as the group started to turn their chairs toward the corner. Bill turned it to CNN, and after a few minutes the TV newscaster appeared on the screen saying, "We interrupt this program to bring you a special report."

CHAPTER 3

A GREAT CALAMITY—UNPREVENTABLE NUCLEAR HOLOCAUST

"This is CNN reporting live from Afghanistan," said the overseas reporter. "At eight o'clock p.m. Eastern US time today, February twenty-eighth, a large mushroom cloud formed over the city of Chebala. There has obviously been a nuclear detonation. At this time it is not known who launched this attack. The death toll will be in the tens of thousands. It is known that this area was a stronghold of Taliban fighters, but there are also US, NATO, and non-NATO military installations in the city. I was in the city yesterday reporting on a local story and can personally verify that there were hundreds, maybe thousands, of US and NATO troops. No one has claimed responsibility," the field reporter said.

Another newscaster interrupted. "This footage has just come in from an amateur videographer." The view showed a nuclear blast seen from a neighboring city about twenty miles away. The detonation looked like the size of a Hiroshima or Nagasaki explosion.

As the newscast continued, various experts voiced their opinions. "This looks like about a twenty-megaton blast, probably detonated at ground level," said a retired US Air Force commander. The newscast continued to offer interviews with local witnesses overlapping with repeated video clips of the explosion. Before long, video clips showed refugees and survivors coming out of the area. People started to ask, "Where is the president?" "What do the Russians have to say?" "Is our nation on tactical alert?" and so on.

A field reporter explained, "There is anguish among all the friends and relatives of the victims. So far many groups are pointing fingers at one another, but no one has come forward to claim responsibility."

ENOUGH QUESTIONS

There was complete silence in the room while the initial reports were coming in.

Manta turned off the TV and recommended a break in the session for them to call their immediate families. He reminded them about their agreement, and told them not to discuss his prediction.

Bill asked everyone to reassemble in the meeting room in half an hour. All of the members of Group T9, with their phones in hand, dispersed out of the room as they were dialing their spouses, children, and parents.

Within thirty minutes the members all slowly started to trickle back into the conference room. When they were all back and had taken their seats, Bill stood up, looked at Manta and said, "Miles, tell us again, how did you know this was going to happen?"

"I told you, I believe I am from the future," said Manta as he looked square into Bill's eyes.

"No," said Bill raising his voice firmly and pointing at the now dark television screen, "tell me how you knew *this*. When? Where? What source?"

"I can't exactly do that," answered Manta, trying to keep the room calm. "Apparently, although I was severely handicapped in my other existence, I was able to learn about historical events. The Chebala incident is one of the most famous conspiracies of all history. Just so you know, there will never be a resolution to this incident. No one will ever step forward. There will be hundreds of different conspiracy theories. I happen to believe that it was a wacked-out scientist, possibly from India or Pakistan, who joined the Jihadist movement. He could have quietly and discreetly built a nuclear bomb with conventional explosives, a collection of conventional detonators, a small foundry, some fissile material, and a personal computer. Remember, Einstein believed the real power of the nuclear weapon was its simplicity. It is possible the bomb was built with smuggled fissile material, or possibly with a new discovery in nonfissile material. Nobody has any evidence because everything and everyone was destroyed, including, probably, the builder of the bomb. It is possible it was set off accidentally.

"Anyway, I was very familiar with this incident because, in the future, it is regarded as one of the most important events in human history — just like December 7, 1941, or 9/11," explained Manta.

"*Where* and *when* did you learn this?" asked Drew skeptically.

"I tried to explain it before. I learned all of my information in a room somewhere with a big TV screen. I lived there and I probably never left," said Manta.

"What period of time and exactly *where* did you live? And mostly, why couldn't you have done something to prevent this tragedy from happening?!" asked Claire, who felt rather exasperated and had tears in her eyes.

Manta paused to take in the emotion in the room. It is a heavy feeling one carries to have just learned about a devastating event, whether a natural disaster or an act of terrorism. Manta had known about this event for years, so he was trying to remain sensitive to the fact that this group—his friends—had just been confronted by the news within the past hour and were still in some degree of shock.

"I did what I could; you have to believe me on that. Over the last ten years I have worked several strategies, including communications to intelligence personnel and private investigations. But you have to understand that there was nothing I could do without possibly making things worse. It is not important at this point to explain my actions—or, rather, my inactions. And I think that I have answered enough questions for now," said Manta.

"Tell us, then, Miles. What *is* important?" asked Carl, somewhat cynically.

"What is important is the information that I have to disseminate to you. I would like to do this over a period of time. I figured three meetings would do it and, like I said before, it is no coincidence that you are all here next week with your families. You have to understand that this day, and the week that follows, have been my whole purpose for the last eight years; I built this group, I included each of you for a reason," said Manta.

"You said you had a religious experience. Are you a priest of some kind? Are we going to have a similar religious experience?" asked Carlos impatiently from the seat beside Bill.

"No, absolutely not," said Manta. "It is essential that our discussions remain purely secular. And, no, I am not a priest. My purpose is not about religious understanding; it is about scientific understanding. Yes, this has been a religious experience for me, but that is my personal journey. As far as what religious significance this has to the world, let the theologians figure that out."

"So, Miles, where, or should I ask, *how* do we start?" asked Rajneesh eagerly.

"Think about what we have seen in the last few decades," said Manta, as he moved to the side of the lectern to address this question. "We saw the world in a downward economic trend, and then, suddenly! The computer changed the whole situation. Then we saw the world in a downward trend again, and, surprise! The Internet changed the economy of the whole world. Housing and real estate boom-and-bust cycles always react to the economic effects of these private business economic game changers. Now governments are threatening default, and people do not see the fundamentals that will pull us out. What we will see in the future is that people will be losing faith in science, and for good reasons. A lot of so-called *science* is becoming like infomercials. Time will reveal that billions of dollars are being spent on backward science. When this is uncovered, it will be like the next banking crisis, where a lot of people had too much control, lined their pockets, and then with their 'exit strategies,' were nowhere to be found.

"But, we have become too dependent on science. Apparently, the social and environmental consequences of letting our scientific efforts hit a dead end are so significant that someone or something of a higher order is interjecting a change, and that is where we come in. Maybe it's something like a bank bailout plan. At least that's my belief or explanation," Manta said in a casual manner.

"Let me explain it this way," said Manta, as he noticed some confused looks. "The reality of the current worldwide situation is that humanity cannot continue to exist without the GEs, the Mitsubishis, the Krupps, or the Hyundais. I do not want to sound like a preacher, but there is only one way to communicate this. In the past, the God of the Bible used worldly governments to affect his purpose. Today, governments are becoming less able to function responsibly to meet the physical needs of the people. Governments have become rendered ineffective by internal paralysis-by-analysis and bipartisan spite. There is too much polarization, undermining efforts, outside influences, and graft. Again, I do not want you to accept my interpretation or beliefs. You can interpret what is happening the way you see it, and let me know later.

"You see," continued Manta, "the corporate structure is more adaptable to change, allowing the entity to maintain usefulness and profitability. Without change you can go extinct. As you well know, despite what a lot of companies advertise, the primary motive of our higher standard of living is not the goodness of the heart. The primary motive is about profit, but, ongoing profit only comes if you have a satisfied customer. The adaptability of the corporate structure is what continues to bring us the goods and services we need, including the clean hospitals and medicines to help cure our ailments.

"In other words," said Manta, "*private enterprise* is going to pull us out of this unsustainable earth-and-humanity-destroying path, with a little help from above. If you are willing, and I know you all well enough to know that you are certainly capable, we are going to create organizations and corporations that will disseminate specific technical information and products that will result in a reversal in the trend of self-destruction by overconsumption."

Manta could tell the group members were finally beginning to understand what he was getting at. He saw many soften their tense faces, which were still in shock of not only learning about his confession, but also about the atomic bomb disaster.

"I want to know if you are available and agreeable to this effort. I would like to ask our chairman to have us vote on whether we will conduct these next three meetings to go over the basics of my disclosures," said Manta, as he finally took his seat at the table. "That concludes my presentation and I apologize I went a few minutes over. I turn the floor back to the chairman."

"Thank you, Miles," said Bill, straightening up in his chair. "Well, normally we would have our question and discussion period at this time, but under the circumstances I think we all want to finish up, get back to our families, and maybe reflect on what we have heard. I know that we could continue on for a long time here, but I suggest that we keep it to a minimum for the time being. I know this is a silly question, but, are there any more questions for Miles?" asked Bill, who scanned the faces of the other members at the table.

"Yes, I have a question," said Marcos, from the left end of the table. "You say that we are going to form this corporation to save the world. How will the public react to this? The public is very wary about the power that large multinational corporations are wielding and the outrageous salaries and bonuses some executives are taking out of the value of their companies. Wouldn't they cry foul about the kind of corporate power you are describing?"

"Well, let me ask you this," responded Manta. "Do you think that Microsoft is going to go away? Or Kawasaki Heavy Industries, BASF, or WD-40? If they do, it is only because a better, cleaner, greener, or more efficient corporation has come along to compete with them for the same market share. Or it could be because they were bought out by a different competing corporation. These companies are too diversified, capitalized, and flat-out depended on to go away. Think about it…Krupp, Siemens, and Mitsubishi built the weapons of war for the losing sides and their whole industries were demolished after World War II. Yet they still survive to be three of the largest employers of the world. If a World War can't kill a corporation of this size and capability, what can?

"The Industrial Revolution has passed a major turning point. With globalization, we are not going to see local manufacturers creating certain products that we need and use every day. Larger corporations know how to develop and deliver these products and raw materials to the markets. More and more, large corporations know that they cannot cause outright harm to people, property, or the environment. Not only is it not worth the cost of fines and loss of image, but also the new generation of leadership in the world will consciously not allow it. Massive population growth and limited resources have made the world conscious of the environmental sensitivity of the planet. Like the tearing down of the Berlin Wall symbolized the end of the Cold War, universal acknowledgement of environmental protection efforts has put an end to ways of the past. Therefore, Marcos, the majority of the public will welcome our efforts by the improvements we deliver to their lives, and to the environment.

"Bill, as I mentioned, I feel that there have been enough questions for now. I would like to take a vote to find out where we stand, and whether or not we will continue to conduct the additional meetings," said Manta.

"You promised to answer one question from each of us about the future, though. I did not get to ask my question," interrupted Walter, feeling somewhat left out.

"Yes, and I will answer your question eventually. Since we ran out of time today, I will open up the question and answer time again when the time is appropriate. But remember, I said I would try to answer your questions in order to prove who I am. I hope the answers and events of today have provided adequate proof. I ask to defer your questions to another time," said Manta.

"OK, what are we supposed to be voting on here again?" asked Walter.

"We are voting on whether or not our group will attend Manta's next three meetings to go over the basics of his disclosures," volunteered Claire.

"I make a motion that we defer the vote until after his next presentation," offered Carlos.

"When would you be ready to present the next phase of your disclosure, Miles?" asked Bill.

"I could do it tomorrow if you wanted me to," said Manta.

"As far as we know, though, the world is mobilizing for nuclear war," brought up Takashi. "I feel compelled to return home."

This comment coming from Takashi, a person who lost family members to a nuclear blast at the end of World War II, resonated with each of the group members. How would those back home be reacting to the crisis?

Rajneesh interjected, "Yes, we will not be going back to work and business as usual on Monday. The stock market will be crashing; air traffic will probably be grounded. Half of the countries in the affected regions will probably go into some degree of martial law. I think about the significant events that have happened here today, and I believe we have to act fast and decisively about both our individual and collective needs. This is not the time to delay, procrastinate, or defer. What I suggest is that we all work tomorrow morning from our homes or our Bay Area offices, send out the appropriate memos, settle the masses, and meet back here tomorrow late afternoon to hear Manta's next disclosure."

"It is going to take me more than half a day to settle the masses," said Katerina logically. "You know we have operations all over the world. Some of my employees may know people were killed in this attack. I am definitely interested in hearing what else Miles has to tell us, but this has also been a lot of information to take in today. Most of us probably won't even be able to sleep through the night. Look, we are heading into the weekend. I make a motion we pick this up on Monday."

There was a pause while all the members contemplated their current positions and were thinking of all they would need to do for their businesses once they got back.

Manta was watching their behavior, mentally digesting how they were going to respond to the next meetings. Deferring the vote to the next meeting was not part of his plan, Plan A anyway. He had multiple backup plans to get to the next phase of his disclosure, including Plan Z where he would fall down at their feet and beg them to hear him through.

It was dawning on all of them that their current obligations to the corporations were remarkably flexible. Although they had always thought of themselves as indispensible, everything was in place for them to walk away. A relatively simple discussion with each of their protégées, and a couple of memos out to the management and general body of the corporation, and they could walk away. Realizing that he was the reason they were in this position, one by one, they all looked over at Manta with devilish grins on their faces.

"Wait a minute," said Rajneesh excitedly. "We have two motions on the table."

"OK, then, let's withdraw the prior motions," said Bill. Organizing a motion, he pronounced, "I make the motion that we all meet here again at eight a.m. Monday morning, hear Miles's next disclosure, and at the end of that meeting have further discussion and a vote on whether or not to continue."

"I second the motion," spoke up Takashi.

"We have a motion and a second, all in favor?" asked Bill to the group.

The voices in the room sounded out "Aye."

"Any objections?" Bill asked, looking around the room and seeing no disagreeing looks. "OK, the ayes have it. See you all Monday at eight."

The members gathered their belongings as well as their thoughts and slowly began dispersing. Manta walked over to Bill as the others were leaving and chatting amongst themselves and asked him for the recording. He pulled the flash card out of the recording device in the front corner of the room and gave it to Manta.

Bill was the last to leave and he shook Miles Manta's hand before doing so. Manta took a brief moment in the silence of the empty room to gather his thoughts as well. He knew many things about the future that were to come, but the deaths from disasters and crises coming to play always were the hardest for him to deal with emotionally. He wished there were a way he could stop bad things from happening or reverse disasters, but he understood the way of the world and also the limitations of his gift. He took one last look out the window at the bay and then closed the blinds. This was the day he had anticipated for years and it had finally happened—he finally revealed his gift to his protégés. He took a deep breath, turned off the lights, left the room, and locked the door. He checked out at the front desk of the conference center and was finally able to think about getting home to be with his family.

CHAPTER 4

A WELCOME RETREAT—HOME

If you are not interested in the private life of Miles Manta, skip this chapter.

It was not a long drive home from the conference center for Miles. The traffic was snarled as usual, but the Golden Gate Bridge looked exceptionally beautiful tonight, causing a soothing effect. The Manta family lived in the town of Tiburon, Marin County, near the north end of the Golden Gate Bridge.

Miles was anxious to get home to see his family. They would no doubt be watching the news about the nuclear blast disaster. He wanted to be there to comfort them and give the reassurance that everything was going to be all right.

He had every confidence that his wife, Lilia, would have their four children settled and comforted with words from the Scriptures, showing the children how the earth was meant to be inhabited forever, and would never be destroyed by a nuclear holocaust (Psalm 37:29; Ecclesiastes 1:4; Matthew 5:5; Revelation 11:18).

On his way home, Miles was thinking about his role as a father. Everyone has a skill set and he felt that being a good father was not in his. He felt badly that he had missed so many family events and school meetings because of business trips and late nights at the office. He wondered whether he was doing enough to give them the personal instruction and education they needed to be happy and protected from the many threats they would face in the rough and tumultuous world.

Work hard, play hard had been his motto for the Manta House. It seemed effective for the most part, but was it too much? Was it appropriate for all the family members?

Maybe his lack of confidence in fatherhood was because he never had a father, at least not before his religious experience.

Miles's spiritual conversion first started when he met the Greek-speaking man at his door. The man's name was Argos and he was the only son and heir of a wealthy Greek shipping merchant. This luxury allowed him to travel all over the world. He became very wise from his experiences. He was also a disciplined man, who had earned a PhD in economics and world history.

Argos had been in a prominent position when a series of disasters struck his life. He lost a child to cancer, then two months later lost his wife and other child in an automobile accident caused by a drunk driver. He refused to believe they were "taken", and he sought answers to his questions about the condition of the dead, unforeseen circumstances, and the resurrection hope described in the Bible (Psalms 146:3,4; Ecclesiastes 9:5; 9:11; John 11:25; Isaiah 26:19). This search led him to a new understanding of the promises contained in the Bible, and that is how he ended up in ministry work, and his eventual knocking on Miles's front door.

Miles started an intense Bible study with Argos, and he thoroughly learned the history contained in the Hebrew and Greek Scriptures. After two years of study, he became a baptized Christian and went into ministry. It was then that he accepted the God of the Bible, Jehovah (יהוה, YHWH), the God of Noah, Abraham, Ishmael, Isaac, and Jesus, as his father (Exodus 6:2–3; Psalm 83:18; Isaiah 12:1–2; 26:4; John 17:6,26). He was also convinced, both from his apparent preexistent knowledge and his religious studies, that this was the same god of Mohamed, who was a direct descendent of Ishmael, Abraham's first born son.

Miles met Lilia while he was active in missionary work. Although she was part of a prominent family in Mexico City, she was living in San Diego, California, doing missionary work as well. Her family ostracized her for her religious beliefs. At the time she and Miles met she was supporting herself by working for a house cleaning company.

Miles and Lilia would describe it as love at first site. They had an old-fashioned courtship, and their wedding was a modest celebration. Within a year they had their first child.

Argos and Miles continued to grow closer through the years. Through the Greek language, Miles and Argos tried to put everything together—who Miles was, why he knew what he knew, and multitudes of other perplexing questions. Argos recommended that the supernatural side of his experience be downplayed, at least until they understood what was going on. Miles made numerous trips to Greece, with and without his family. In the effort to avert prying into his language background he told others that he had learned fluent Greek through home study courses, travel experience, and Argos's excellent teaching ability

Argos did not want to push any ideas that Miles was something supernatural or a prophet, and he highly discouraged Miles from making such presumptions. In the end, Argos could only deduce that if there was an "inspiration" that it was from Satan the Devil. This led to disagreements, but they agreed to disagree, as neither of them had a complete understanding of the situation. They both tried to figure out what he should do with his gift.

Argos was a very wise and experienced man. His position was solidly based on Jesus's example that true Christians are not to be involved with "the World" (2 Corinthians 4:4; 1 John 5:19; Revelation 12:9; John 15:19, 18:36; James 4:4).

Miles argued that Abraham was successful negotiating with God to spare Sodom & Gomorrah if he could find just one righteous person. Another example was Jonah, where he successfully, though reluctantly, convinced the population of Nineveh to repent, and saved them from destruction.

Miles felt that "private enterprise" was not part of the blasphemous political systems that the Bible shows to be destroyed in the prophesies of Daniel's image, and the wild beast of Revelation (Daniel chapters 2 &3; Revelation 13:1–18, 19:19–21). He also felt that private enterprise was not part of Babylon the Great, and the "merchants" only wept at her demise (Revelation 17:1–6, 18:11–16).

He argued that the apostles had day jobs, and some were even wealthy, at least when they started their ministry. Four were definitely fisherman, and Matthew was apparently a wealthy tax collector (Matthew 4:18–21; 9:9–13). Luke was apparently an educated physician (Luke 4:38; Acts 28:8). Jesus himself was a carpenter by trade. Though not part of the original twelve apostles, Paul was a tent maker (Acts 18:3).

Argos wanted to err on the side of caution in order to protect the congregation, and he believed that Miles's situation could cause people to stumble in their faith if they found out about it. Argos and Miles discussed it and mutually agreed to part ways. Miles left the congregation. This coincided with the Manta family moving to a new area across the Bay.

Miles had anticipated that Argos would eventually be gone from his life. Maybe they parted ways because of a self-fulfilling prophecy—Miles expected it, so it came to be. Like many of his other relationships, Miles again had become a threat. Both men knew that no committee could resolve the growing conflict between them.

Again an outcast, Miles thought about the life of Mahatmas Gandhi and his comments about Jesus's Sermon on the Mount. Gandhi believed that the whole world would be at peace if people just followed the direction provided in these simple verses of Matthew Chapters 5 through 7. In particular, Miles was comforted by Jesus's words at Matthew 6:5–8, which gave him the peace of mind that he never had to feel bad about not fitting in with society.

Although Miles had confided in Argos in many ways, he kept his street smarts. Miles did not share his business life with him.

He didn't disclose everything to Lilia either, at least not everything about his business. He did share with her the apparent supernatural experiences, including the pending nuclear disaster. She agreed not to disclose any information, and had no problems carrying on as if she had no special knowledge or that the Manta family was any different from any other family.

Lilia did recognize that Miles was different, especially when the bank account started growing exponentially. The money did not mean anything to her; it was just her job to manage its whereabouts. She maintained the household's low profile and made sure they lived well within their means. She loved Miles more than anything, and she would keep his secrets to the very end.

Regarding Lilia, Miles was sure of one thing: how good she felt in his arms. And he was anxious to see her as he pulled into the driveway and ran up to the house.

Lilia greeted him in the hall between the entry and the kitchen. They fell into each other's arms, holding each other for a period of time as they breathed deeply. The long embrace was like a melting sensation for Miles—a safe haven where nothing else mattered. The world could be vaporized around them and neither one would have cared.

"How did your big day go?" asked Lilia.

"Great, we are meeting again on Monday," he replied. "How are the kids?" whispered Miles.

"They are in the family room," she replied. "Papa's home!" she shouted as the two of them walked toward where the children were seated on the floor.

The youngest son and daughter raced to jump into their father's arms. The older two boys just nodded and mumbled, "Hey, what's up?" without moving from the floor. Miles bent over and gave them each a pat on the back and a fist pump.

The oldest son, John, a fifteen-year-old, asked, "Dad, do you know what's going on?!"

"Yes, we were watching the news reports during our Trek group meeting after the explosion happened," replied Miles. "What do you think about the whole thing?"

The three sons and their father sat on the couch and engaged in conversation about the unfolding events. As the boys discussed the news, the youngest, Rosaria, an eight-year-old girl, was helping her mother in the kitchen, where she loved spending time with her mother when the boys were not around.

The oldest son John explained about how they had just been studying the history of Afghanistan in school and he could see how the Russians could be involved, because they would want to get control of the area. The youngest boy, Mitchell, a ten-year-old, didn't know how to comprehend the significance of the disaster. Miles could tell he felt scared but didn't want to admit it.

The second to the oldest son, Richard, who was a very smart boy, was much more reserved in expressing his opinions, especially around his older, dominating brother. But Miles could tell he was taking all the information in like a sponge.

"Dinner's ready!" yelled Rosaria, and they all moved into the dining room.

"Ah, smells great. Chili verde, my favorite," said Miles, smiling and leaning over his plate as he took in a deep whiff through his nose. He led his family in prayer before they started eating.

During dinner the family engaged in more conversation about the world events, and how they would affect the Manta family.

"I will assure you," said Miles. "The event that happened today will never be forgotten. It will go down in the annals of history."

Lilia's chili verde was excellent, as always, and Miles scraped his plate clean. They all took their plates to the kitchen when they were done. After dinner, the children each went to their bedrooms and began getting ready for bed. Having dinner later in the evening was not unusual because it was often the only way they could all eat together as a family.

Before the children went to bed, Lilia read Rosaria a story. Miles read all the boys a poem on the couch in the family room. It was something he had wrote while away on a business trip when he was unable to sleep and thinking about his family.

Good Fathers and Lonely Discoverers
> There are a lot of good fathers
> They play baseball
> Join boy scouts
> Are part of social circles
> Sometimes good-looking
> Discoverers are not so fortunate
> Their mission is lonely
> Not conducive to groups
> Fear of failure
> Not so attractive
> But the boys should know
> They can be self-made
> And that is the best thing
> A father can do for his sons
> I am satisfied
> I have enabled
> You to be so

He then told the boys that tomorrow was their day. They could do whatever they wanted to do, as long as it did not involve video games. They pretended they were excited and hugged their dad goodnight.

Miles leaned back against the couch and felt exhausted. He had been in this state many times before, but this time it was different. This time his fatigue was accompanied by a sense of freedom, as if he had just been rescued from captivity. He had finally let his Trek associates in on his knowledge. He no longer felt burdened by having to keep it in. With a deep breath, he got up and made his way down the hall to their bedroom. Lilia had the bed waiting. He kissed her forehead and within seconds after his head hit the pillow, he fell sound asleep.

When Miles awoke, he felt as if he had just closed his eyes for a second. It was clearly daytime. He immediately felt fully awake. It was such a strange feeling, like he had been cleansed somehow. He had a complete sense of rejuvenation. Everything he saw had a different look—even the air itself seemed strangely new. He glanced at the clock; it was after three o'clock in the afternoon. He had been asleep for about sixteen hours!

He was so relaxed he felt numb. But just lying in bed did not suit him. He sat up, grabbed a pen and a notepad from his nightstand, and wrote what came to his mind:

The Promise
How far you have strayed
Now you need to be saved
Saved from yourself
Saved from your self-inflicted destruction
You should be ashamed, but you're not
You only look to pass blame
Yes, I regretted making you
Saved by three,
Now your science proves you are brothers
But in a second you abandon the facts
When they don't support your case
Just like you abandon me
For whatever suits you
Remember—you were naked when I made you
And you were not ashamed
And it was good, until you shamed yourself
Now, you are far removed
Tricked by simple lies

Lies are never simple
They will always inflict pain and suffering
You were first animals
But I gave you more than instincts
In our image, I made you
You cannot discern all
The Spirit is immaterial
Your understanding is limited
You think others deliver the curse
You bring curses upon yourselves
What is the difference—a curse is a curse
If you curse yourself
You can also bless yourself
Look, your brother needs help
Hasn't he suffered enough?
Haven't you all suffered enough?
You have come to have great power
You are no smarter than those that preceded you
You benefit from their advances
Don't think that you have better justice
Study the Law—you will see
Now you think your power is great
But only the best of you know the limits
And few understand the real cost
Those few are squashed by the loud and obnoxious
They know nothing, except how to exploit the formula
Get trust, get control, all for riches
Your efforts are in vain
Your time is limited
There will be a new beginning
And my involvement is eminent

Miles just let the words flow off the end of the pen. They seemed to come from his subconscious and had only a vague meaning to him.

He felt so light and free after writing his poem. He felt more refreshed than he could ever remember. It felt good. He was starting to realize the more he exposed his gift, the lesser his burden.

He put down his pen and notepad and got up to finally start the day and look for his family. He was excited to spend some quality time with his family and was eager to find out what they wanted to do. They ended up spending the day around the house playing games and watching videos.

On Sunday morning the family attended a public discourse at the local congregation. Both in the opening prayer and before the guest speaker was introduced the Presiding Overseer highlighted the Chebala disaster.

The Presiding Overseer was an older gentleman who had observed many calamitous events over the decades, and he advised the congregation not to use the incident to make any presumptuous claims that Armageddon was upon them. He described that even though they teach that they are in the last days of mankind's attempt at self-rule (Genesis 3:20–24), and they should always remain vigilant as if the end is tomorrow (Luke 21:34–36), only God knows the exact day and hour when the end would come (Matthew 24:36–38). Though there will be signs (2 Timothy 3:1–5; Matthew 24:3–14), there will be no fanfare or warning when the end occurs, and it will come as a thief in the night (1 Thessalonians 5:2).

The public discourse was entitled "Are Science and Religion Compatible?" The speaker described that although the Bible is not a science book, it's collection of books dating back about 3,700 years are remarkably free of the outlandish beliefs that prevailed amongst contemporary teachings, including mythological tales of an afterlife, worship of animal images, and the mistaken belief that the earth was the center of the solar system.

The speaker described how the biblical creation account is misrepresented by those who claim the universe was made in six literal twenty-four-hour days. The speaker referred to the first sentence of the Bible, Genesis 1:1, that shows how the "heavens and the earth" existed before the creative "days" started, therefore, from the perspective of the biblical account of creation, the universe could have existed for billions and billions of years before the "creative days" began. He described how the Bible has variations in the use of the word "day", for example, his instructions to Adam regarding the tree of knowledge of good and bad, God says "for in the 'day' that you eat from it you will positively die" (Genesis 2:17), then Adam lived to be nine hundred and thirty years old (Genesis 5:5), obviously not a literal twenty-four-hour day. Also, the example that a creative day is not a literal day because God is still in his rest "day", about six thousand years after the sixth creative day (Genesis 2:2, Psalm 95:11, Hebrews 4:1–11).

The speaker provided evidence of a sudden and catastrophic worldwide flood, including widely dispersed marine fossils observed in sedimentary layers across the globe, frozen mammoth discoveries in Siberia, and the four-hundred-foot elevation changes of the Bahamas's Blue Holes. The scientific data indicate that these events had occurred only thousands of years ago, not the millions and millions of years often recklessly declared by the scientific community.

The speaker provided examples of how the science of anthropology shows that we descended from three racial types, caucasoid, africoid, and mongloid, corresponding to the three families of the sons of Noah, and that the human population of the Earth came out of the area commonly referred to as Mesopotamia. Examples were provided showing how a belief in any connection between modern humans and bone fragments of apes would require a huge amount of blind faith, and that stories of pre-human Neanderthals, could easily be ancient ordinary humans that found residence in caves, similar to civilizations that existed in the Americas only several hundred years ago.

The speaker provided examples of how there can be fraud in science, including Piltdown Man, and the exaggeration of what we know about the past. He gave an example of how modern science cannot determine who lived at Qumran only two thousand years ago, the city that existed near where the Dead Sea Scrolls were discovered in 1946, yet some scientists will try to tell you exactly what was happening in a remote swamp four million years ago.

The speaker described how radiocarbon dating is subject to environmental conditions, such as the "marine effect" and "hard water effect", and the synergistic effects of the countless combinations and permutations of environmental factors. Radiocarbon dating has limits, for example, a search of the literature will show radiocarbon dating is generally limited to dating samples no more than 50,000 years old, because samples older than that have insufficient Carbon-14. Therefore, radiocarbon dating cannot be used to disprove creation, because Genesis 1:1 indicates no limit to age of "heavens and earth" (billions and billions of years is fine), and the "creative days" could well be tens of thousands of years in length.

The speaker told a story how he was in line to see a science exhibit at a prominent museum and two people in line with him were discussing a recent discovery that the design shape of a whale's fin produced the most efficient wind turbine blade. One of the persons said, "That's what millions and millions of years of evolution can do." The speaker pointed out to the audience that this explanation has no scientific basis, and is no different from the "God can do anything" explanation often used by religious zealots.

He provided an example where he asked them to imagine taking apart a simple wood chair and dispersing its parts throughout a large lake. What is the probability that this chair would reassemble itself with a new finished coating? The laws of probability could calculate a number, but what does it mean? It would be such a high number that the reality of it happening would be zero. Now, imagine the chain of events that would result in the design and development of the human eye—an auto-focusing, full-color, stereographic, self-healing camera system with matching reflexive covers, connected to an ongoing memory of yet unknown capacity, that forms itself from an embryo of two cells, can occur with the look of Greta Garbo, and grows to full size within the first three months of birth remaining exactly the same size for life. Current population data show this process has repeated itself over seven billion times, and over history has repeated itself tens of billions of times over all of the known human history, without any known deviation from the original pair. Now apply this to every living animal that has eyes. Is it statistically possible for the complex systems associated with vision to occur by chance? If you believe this is possible, statistical analysis would show your belief is based more on faith than science.

The laws of statistics show the probability of successive events decreases exponentially, for example, when using a six-sided die, the chance of rolling two "ones" in a row is $1/6^2$, or $1/36$, and the chance of rolling three ones in a row is $1/6^3$, or $1/216$. Therefore, the probability of rolling successive "ones" with a six-sided die is given by the equation $1/6^x$. He rhetorically asked the audience what would be the value of x for our simple chair example, then suggested a modest value of x = 50 successive events, resulting in a probability of 8 followed by thirty-eight zeros. It would take a lot of faith to believe the chair would rebuild itself by chance, and this does not even consider that we started the experiment with a fully designed and manufactured chair.

The speaker continued with the theme of the human eye, explaining that the first day of creation when God said "Let there be light" was not referring to light, the electromagnetic wave, but to the advent of vision, the ability to see light. This is similar to the quandary purported by the question – if a tree falls in the woods does it make a sound? Not if sound is defined by the ability to hear. Prior to the first day of creation there was only a spiritual and nonphysical reality to living beings. What makes the physical world, first and foremost, is vision, the ability to see. The human brain, which dictates our existence, stems from the great mechanism of vision. No doubt, the advent of vision made the angels shout out in applause (Job 38:7).

Continuing with more scientifically based logic, the speaker described how according to the laws of thermodynamics, all systems progress from order to disorder, therefore, "millions and millions" of years would result is less probability of complex systems being formed by chance.

He continued his story about visiting the museum and referred to a display that claimed to show an example of evolution. The display told a story about how a drought had killed off a particular food source from a group of finches, and that only finches with a certain beak geometry survived. The speaker explained that this observation is not evolution, but an example of adaptation of an existing species. The speaker explained that this "example of evolution" is no different, and rather pales in comparison, to the process of using dominant and recessive traits mastered by the long line of dog breeders that derived all of our known breeds of domesticated dogs from the then existing stock. The main point the speaker emphasized was that the variety of traits is from the genetic makeup contained in the original pair of birds, and there is not a new bird species—they are still finches.

The speaker opined that the belief of complex designs coming from random events has no known scientific or mathematical basis. The inescapable paradox that applies to any model of "order from disorder" is that the model itself has an element of design, as shown in our model of the chair in the ocean. The only way to add a semblance of science to the problem is to add a fudge factor, such as "billions and billions" of years, as an attempt to validate whimsical notions about the origin of life. Then, in brilliant human fashion, elitist "nonprofit" organizations are organized to garner billions and billions of dollars to conduct research for evidence to search for proof—essentially, proof that God does not exist. The speaker remarked that this type of research always concludes with the recommendation that additional research is required, preserving an ongoing revenue stream.

The speaker concluded his story by noting that during his museum visit he observed the typical misrepresentations of the survival of the fittest, natural selection, and extinction, all twisted into dogmatic "proofs" of evolution of something from nothing. All of these processes are not proofs, but are natural occurrences, and depend on an original sample of genetic traits from an original pair. Science tells us that these traits were part of the original DNA. The speaker admitted that he was a formally trained scientist who despite efforts has never observed a fossil record showing one species evolving to another.

Comments were made about the cause of the perceived division between the belief in God and the higher levels of science, and that a belief in God can accentuate your passion and understanding of the natural and physical sciences. Examples were provided of two of the greatest physical scientists of all time, Isaac Newton and Michael Faraday, who were staunch believers in God and that the Bible is God's inspired word. Obviously, a belief in God did not hold them back, and they were inspired to advance further in their appreciation of the laws of science.

He explained that much of modern day science is tainted with the lust for movie-star status, money, power, and control of people, in the exact same manner as many organized religions. The speaker described a common trait between organized religion and the world's scientific institutions of "groupthink" psychology, where everyone in the group believes it's true because they want control or acceptance, and they do not question the underlying merit of the belief for fear they seem stupid or cast out.

The speaker concluded with comments about the reliability of the Bible (2 Timothy 3:16), the legal meaning of the sacrifice made by Jesus Christ (Deuteronomy 19:21; 1 John 4:9–10), and the wonderful science that people will have an eternity to learn about when God's Kingdom reigns over the earth (Genesis 2:19–20; Revelation 21:1–4).

The speaker received a warm applause. It was apparent that most of the audience appreciated the information and felt inspired to stay spiritually strong in the face of uncertain times (Proverbs 18:10). The group concluded with a song and a prayer.

After the meeting, the Manta family enjoyed a barbeque lunch with friends, and then made their way home.

Later in the evening, after the children were situated for the night, Miles and Lilia got ready for bed.

Over the years Lilia had made photo albums in the form of story boards showing many of the experiences the family had shared together. These albums would include a sequence of pictures with titles of the places and things they experienced.

On one occasion she was using Miles's old Bible at a congregation meeting and hidden in between the pages she found a piece of paper with a poem.

Miles had written the poem soon after he met Argos. In fact, the Bible was a gift from Argos. It was clad with a worn out leather cover with his embossed name barely visible on the front cover.

She never told Miles that she had found it, and she wanted to save it for a special moment.

Before they went to bed, Lilia told Miles the story about how she found it, and pulling out the old piece of paper from an envelope hidden in a photo album, she gave it to him. It read as follows:

In Search of Eve
I can imagine

Seeing her
Feeling her
Hearing her voice

Beside me
Within reach
Anytime
Caressing her curves

Her fingers
Message my head
Cleansing me
No accounting

Keen senses
She benefits
The rarest trust
Completely shared

Longing for her
Every time I go
Never grows old
Infinite space to grow

Could I be
The happiest man
Ever to have lived?

After she read it to him she asked, "I hope I have fulfilled your search?"

He told her that she had in every way. He pulled the covers over them, and as the blankets were deflating and settling over their naked bodies he said to her, "I don't deserve you".

CHAPTER 5

PRIVATE ENTERPRISE TO SAVE THE WORLD

Everything was unfolding just as Manta had described. Just like 9/11, the world airlines were temporarily grounded and it seemed like the perceived normal way of life was forever changed. The worldwide stock markets were crashing. People hovered around their news programs listening to the so-called experts decipher the extreme points of view, opinions, and finger-pointing.

To the surprise of many television viewers, there were great examples of unity from the major nations. Showing strong leadership and cooperation, they quickly arranged a United Nations meeting, and world leaders spoke with one another and demonstrated to the world a collective effort to deter any escalation in tensions or conflict. Aid packages were developed for the people of Afghanistan and were rushed to the affected thousands.

Still, the title or subheading to practically every news program was, "Who did it?" Who would have been capable of delivering and igniting a nuclear device in the middle of Afghanistan? What would someone or some group's motive be to do so? Although numerous conspiracies were circulating, nothing was conclusive. All the best reporters could do was "bring up the usual suspects" in the media newsrooms for more discussion and debate.

The Taliban and Al-Qaeda, the Iranians or other fundamentalist Islamic groups, radical Israeli groups, the Russians, the Pakistanis, North Koreans, and everyone else any one could imagine were brought into the fray.

Still, as Manta predicted, no person or group claimed responsibility. The United Nations summoned a commission to investigate the matter further. Many who witnessed the aftermath of the JFK assassination compared it to the current ongoing debates and conspiracy theories.

It has been said, calamities change the minds of men. If there was any bright side to this tragedy, it was a unifying effort to ban the world of nuclear weapons. More and more freelance journalism and Internet blogs were arguing how just a small operator could have made this device. Regardless of who did it, no one knew what the intentions were.

TREK MEETING NUMBER ONE

A feeling of eerie apprehension prevailed over each of the T9 group members as they commuted in for the Monday morning meeting.

Some of them spent time over the weekend trying to figure out a logical explanation of how Manta had predicted the event, and tried to decipher a flaw in his story that they could use to pin him down. Others had conversed about the possibility that Manta was a terrorist, and at what point should they alert authorities.

Still others with more religious or spiritual dispositions studied the Scriptural references that he had alluded to, and maintained a more alert stature during their Sunday church services. Due to the circumstances, people of all faiths were praying more in earnest. The pews of all different types of churches began to fill up after the devastating event occurred.

Members of Group T9 filed back into the same conference room used the prior Friday, chatting with each other and helping themselves to the scones ,coffee, and tea Manta arranged for the group this morning. Eventually they all sat down at the table in whichever seat they wished.

Manta walked over to the window to open the blinds and took his place at the front of the room behind the lectern. "I hope all of you had a pleasant weekend. I can imagine most of you had at least one hard-pillow night. Knowing your backgrounds as well as I do, hard-pillow nights should not be anything new for any of you," said Manta, eager to begin the first of his three-series disclosures. Manta continued right on. "We have always kept the small talk to a minimum during our regular meetings and for the sake of the best use of our time and expediency, I hope that trend continues. I call this meeting to order.

"Since this is a continuation of last week's meeting, I would like to turn over the chairmanship to Mr. Bill Oliver again," said Manta. Bill switched seats with Walter as Manta continued to talk, so he could have the front right hand seat where they always placed the chairman. "In fact, I would like to suggest that the series of these three meetings this week all be chaired by Bill. How does everyone feel about that?" asked Manta as he scanned the room for feedback.

The entire group was in agreement. Over the years, Bill Oliver had gained respect from all of the group members as an experienced, levelheaded, highly diplomatic, and naturally charismatic leader.

"Thank you. Glad to oblige," said Bill, feeling honored to hold this position again in this meeting and the next two. "OK, let's get started. We left off last week after we heard Miles's far-fetched claim that he is some type of inspired prophet that has come back in time to save the world. He has turned all of our lives upside down by revealing that over the last eight years we have all been groomed by him to share in this heroic effort of using our business acumen to ride our corporate chariots of Inc, Limited, Gmbh, and Spa, into the sunset to save the world. With an ultimate arm-twisting gesture, he predicted the world's most destructive peacetime genocide when over twenty thousand people were wiped off the face of Earth in a single moment on Friday evening.

"We voted unanimously to hear Miles's first disclosure today, at the end of which we will vote to determine if we want to stay involved," continued Bill. "And I'm sure Miles will explain more about what's involved if we vote yes to stay involved. The question will be, what if we vote not to stay involved, either individually or as a group?"

"I have thought about that also," Rajneesh added right away from across the table from Bill. "If the vote is not unanimous, then does that mean the whole group stops, or just the individuals that vote against staying involved? I would suspect that the individuals who vote against staying involved would leave the group and promise not to divulge what we have discussed, right?"

"Let's do this," suggested Bill. "Let's take the vote at the end of the meeting today. If the vote is not unanimous, then we will have a discussion and a second vote regarding how to split up the group at that time."

"I agree," spoke up Carl from farther down the table. "This whole effort today will come down to a confidence vote in Miles. If we vote unanimously, then we have full confidence. If he gets a majority, then he will likely keep his group and those that vote in favor, and the others will leave. If he gets a minority, then that will reflect a serious lack of confidence and our second vote may be a vote on his sanity."

A few chuckles were heard muffled under members' breaths. No one from the group had shed light to any of the other members whether or not they felt like they were in support of Miles or not, so, at this point in the meeting, they could only guess what the other members were really thinking.

"I make a motion we take the first vote at the end of today's meeting. Then we can have further discussion about the second vote if it is even necessary," said Drew, wanting to move on to Miles's presentation.

"I second it," said Claire.

"We have a motion and a second. All in favor?" asked Bill.

"Aye, aye, aye," said the group members from around the table.

"All opposed? None opposed. The motion carries."

Miles was standing quietly throughout this whole discussion at the front of the room. He had a high level of confidence that he would get the support of his fellow Trek group members, but they certainly weren't obligated, and he knew they had the right to leave the group if they so desired.

"OK, Mr. Manta. You have the floor. Take us away. I mean that figuratively, of course. You are not allowed to transport us anywhere or hijack us into your space-ship," said Bill jokingly. A few members scoffed at this remark.

"Very funny," said Manta. "That is actually a good lead-in to what I want to tell you about first. I know all of you are serious types…well, with the exception of Bill. And we all know how much we appreciate that. You have to understand that in my situation there are some things I cannot talk about. I could tell you things that would scare the living daylights out of you. Things are being done today—right now, as we speak—that are very wrong and unacceptable. The world—the environment—is very resilient, but when it gets pushed to a certain point, things can change very quickly. You are all familiar with exponential growth. Bacteria in a sealed bottle doubles in quantity at a compounding rate until the final hour when they suddenly die from their own waste. That is the situation the world could find itself in if changes are not made.

"Yes, Bill was correct when he referred to the corporate chariots riding off into the sunset to save the world," said Manta. "As it turns out, only private enterprise can save the world. We should all sense this. Every day we see examples of how government-based programs quickly become bogged down by bureaucracy, where every angle is manipulated by profiteers or self-serving forces of job preservation to produce maximum inefficiency and the highest cost. Government programs only work when the lowest common denominators are people like Neal Armstrong and Gene Kranz. Yes, the Apollo rockets were built by the lowest bidder, but in this case they were also the best in the world. These guys were raised on farms where, as children, every morning they woke up trying to figure out a clever new stunt that put their life in danger. They learned by experience how to handle taking risks. You can't legislate this kind of aptitude or learn it from a book. Yes, of course, government has its place. And yes, regulation is required. But, surprisingly, it is corporations themselves that are the only ones that can make regulations successful. The good intentions of the regulations have to be what the corporations want to advertise: "We do this because we care." Also, it is the legitimate corporations that have the proper perspective of taking risks and the consequences of failure. Governments should not be involved in the business of taking risks, or absorbing the losses of corporations that made wrong choices. Corporations must adapt to change and innovation, and if they cannot, they need to go away or get out of the way.

"I apologize.... I am already going off on a tangent," said Manta, changing positions and shuffling around a few of his notes to regain his focus. "I know all of your situations. I know your businesses, I know your families, and I know your strengths and weaknesses. I would never ask anything from you that I felt you were not capable of. I would never intentionally jeopardize your health and safety, or that of your families. As I have said before, I am personally taking financial responsibility that your companies will not suffer from your absence. Mainly, I have already taken care of that through the placement of protégés and forcing you to orchestrate your exit strategies, as all of you now realize," Manta explained.

"I am asking you to sit through a series of meetings which I want to call 'lessons,'" said Manta. "The subject matter will be mostly, if not entirely, scientific in nature, and you will understand why this is the case as we progress. In addition to the three lessons, I will assign you a couple of reading assignments. It is important for you to read and study this information.

"I know that some of you are not technical people," continued Manta. "You might be intimidated by some of the science, so I apologize in advance. All I can do is spoon-feed the information to you in bits and pieces. We are all educated adults here, and what I have to describe to you should not be that difficult for you to understand. I have faith in you. It should be no more complicated than what kids are learning in high school today. My material is written with a historical perspective. And it is intended to highlight some of the greatest discoveries in science, some Ys in the road, and some wrong turns. You will have plenty of opportunity to go over the handouts and conduct text book and internet searches about the historical significance of the mathematical concepts, including the controversies about the area-under-the-curve and instantaneous-velocity.

"From this information we will put our intellectual property in place," explained Manta. "Corporate structures will be set up on a worldwide basis to fairly disseminate the use of our technology for the benefit of the public and the environment. There will be tremendous costs both in the materials and human resources necessary to make this happen. This is a for-profit venture, and we are not interested in government financial assistance.

"An important goal for our enterprise is to clean up the bad image that has been portrayed of corporate executives," said Manta. "We all know the damage that has been done by several sectors, mainly by the banking sector, where unbridled greed and easy money has destroyed the image of corporate executives. We can only do this by example.

I know what kind of CEOs you are not. You are not the type that has mastered the art of paying yourself extravagant bonuses for doing nothing. I know you would not pay yourselves a bonus if your company lost money. I know how all of you have lived and conducted your businesses. I have seen how you have passed the baton, and how you have encouraged and motivated others. Also, we have plenty of good examples outside our ranks in many different countries and cultures."

Manta moved to the side of the lectern and continued. "A common belief that I know we all share is the great potential human beings have to design and build!" exclaimed Manta. "We truly can do anything we put our minds to. We have witnessed great technological advancements in short periods of time. But we got a little ahead of ourselves. There was something lost. The key to getting back on track lies in happy, productive people and families, and good leadership—leadership that is guided by fundamentally good ideas and ethics.

"You, my friends, will play a great part in that endeavor," said Manta. "Remember how I described the Ys in the road that were taken in the recent past? Well, there will be many more Ys in the immediate future. I am certain that you are the most qualified group ever assembled in the history of humankind to make the best decisions. Do not think that I have all the answers, though. We will learn a lot as we go and we are in this together.

"What I have to present to you will give you the fundamental tools of understanding in order to learn how we can best shape the future," said Manta. "So put on your thinking caps, and let's talk about the biggest breakthrough in the history of humankind."

CHAPTER 6

THE FUNDAMENTAL NATURE OF THE UNIVERSE

Some members readjusted themselves in their seats because they knew he was about to get into heavier material. They sat up straighter and had their eyes on Manta as he organized some papers on the lectern.

He cleared his throat and began.

"Since the dawn of time, major scientific and technical discoveries have changed the course of human history. These discoveries range from smaller mechanical devices, such as the sewing machine, to more complex contraptions, such as the Bessemer converter, to the incredibly complicated science involved with atomic energy.

"In each case, the inventors of these game-changing discoveries had to fear for their lives. Why?" Manta asked the group rhetorically. "From the low-paid worker to wealthy owners of companies, in each case the sovereignty and livelihoods of groups of people were threatened by these new findings.

"I want to warn you all that the information I am presenting and we will be discussing will cause some people to become angry. They will feel hostility toward us and try to discredit the information we present. The opposing side will be strong and controlling, but we must be strong and understand that what we are doing is the only option for the long run. I know that you are mature and experienced enough to hear this warning.

"The first thing we are going to talk about today will someday be universally known as the Fundamental Nature of the Universe. All other scientific laws, phenomena, explanations, and observations are based on this one basic explanation of the state of our universe."

THE SPIROGRAPHIC GRID

Manta leaned over the podium to get the full attention of the group.

"Right now, as we all sit here," Manta continued, "it is a scientific fact that we are traveling around the sun at a velocity of about 108,000 km/hr (67,100 mph). In addition to traveling at this great speed, we are constantly changing directions.

"When I say we are constantly changing directions, I am referring to us being on the surface of the earth, which is rotating with a surface velocity of about 1,670 km/hr (1,040 mph) at the equator. Therefore, we are constantly switching directions as we spin and orbit the sun.

"In addition, we know that our solar system is part of the Milky Way Galaxy, which has a rotational pattern that adds to our constant movement and directional changes.

"Looking out into the heavens, we have learned that there are billions and billions of other galaxies and that our own Milky Way Galaxy must be rotating around a multitude of different galaxies in a myriad of directional combinations and permutations."

The group seemed to be with him so far, since this was basic scientific material. "It will be discovered in the future that the order of these motions is not random," Manta explained. "The pattern of these motions is actually the most precise arrangement of movement ever imaginable.

"This order of motion is called the Spirographic Grid, and examples are shown here as Figure 1," Manta explained as he used his laser pointer to highlight Figure 1 of his multimedia presentation that was projecting images on a screen. He also passed around a handout showing various spirographic images and other images included in his multimedia presentation to everyone at the table.

The members all looked quizzically at the sheets of paper in front of them and felt like they were looking at a frozen image of a kaleidoscope. Some recalled the popular toy that many grade school children played with that included various wheels with gears used to make spirographic images.

Manta continued on once everyone had a copy of the handout. "The discovery of this ordered motion of the universe has profound consequences for the understanding of natural phenomena. This spirographic motion, and the motion and force equations that stem from it, are referred to as the Fundamental Nature of the Universe, which is important to remember.

"Another significant aspect of this highly precise Spirographic Grid pattern is the extremely high velocity of the motion. For example, it takes a tiny fraction of a microsecond of time for our entire solar system to travel the entire cycle of the entire Spirographic Grid pattern. This cyclic motion occurs at a resonating frequency in the order of the velocity of the speed of light.

"Are you with me so far?" Manta asked the group as he looked around, and could tell they were definitely interested in where he was going with this.

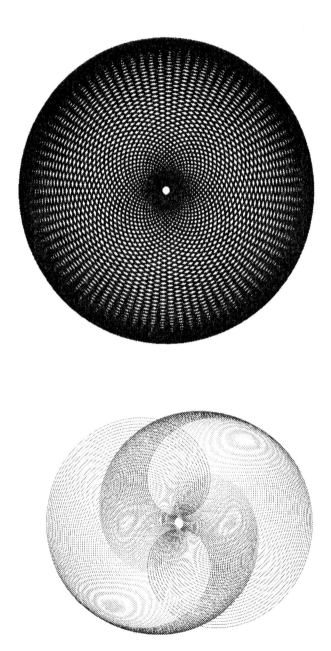

Figure 1- Examples of 2D Spirographic Patterns. The Spirographic motion of the universe is three dimensional (3D) motion and has true left-hand bias. See Chapter 10 and www.wikappendix.com regarding right-hand rule of electrical induction.

DYNAMIC VERSES CORPOREAL MOTION

"Now that the Fundamental Nature of the Universe, which I'll call the FNU, has been described, I will clarify what is meant by the word *motion*. A clarification of our use of the term 'motion' is important when you consider that Albert Einstein formed a new religion when he made 'relative motion' the principal theme of his 'relativity' doctrine.

"First of all, relative motion is an elementary part of the study of mechanics, including statics and dynamics, studied by all engineering students as part of their education curriculum.

For the purpose of understanding the Spirographic Grid, we need to clarify two important types of motion, which are 'dynamic' and 'corporeal.'

"Dynamic motion is the combination of translational motion and rotational motion of any mass. Translational motion is the line of travel followed by the center-of-mass of the object, and rotational motion is simply the velocity of rotation. A baseball, for example, will have a translational motion along the path it travels from the pitcher to the catcher, and it will also have a rotational motion, except when the pitcher throws a 'knuckle ball' where he intentionally throws it with no spin.

"Dynamic motion is very well understood, and engineers can model and calculate the velocities and forces associated with this motion. One example of this design capability is the advent and perfection of the connecting rod used in reciprocating engines. Remember, during the Industrial Revolution, steam locomotives were made that traveled over one hundred and fifty kilometers per hour. Can you imagine the forces associated with a massive, multi-ton rod that connected the train wheel to the piston flying around at thousands of revolutions per minute? To make it work, the designers created counter weights to perfectly balance the massive forces caused by the translational and rotational forces of this large rotating mass. Otherwise, the whole connecting rod assembly would have caused such severe vibrations that it would have torn the locomotive into pieces.

"The dynamic motion of a connecting rod is actually rather complex; it involves a combination of translational—or linear—velocity, and angular—or rotational—velocity. Each end is accelerating to a top speed and then decelerating to zero velocity as it turns back in the opposite direction—the whole time rocking with an angular rotation. It is a wonder that scientists and engineers can model this complex motion with mathematical equations. This deep understanding of relative motion is what makes it possible to produce reciprocating race car engines that can run smooth as silk at astonishing rates of over 20,000 rpm."

Manta flipped to the next page of his notes on the podium and continued.

"The other type of motion that we need to understand is called 'corporeal motion,' which is the motion of a mass relative to a real, but nearly mass-less, media. To benefit our discussion, we can think of this media as a 'submass' for now. Please note that corporeal motion is a *relative* motion; therefore, a mass, such as a planet, could be at rest while the submass media moves, or, the submass media could be at rest while the planet moves.

"An illustration of corporeal motion is the relative motion between a fish and water, for example, where the fish is swimming against a strong current and could actually be moving backward relative to the ground. In this example, the fish is the mass and the water is the submass.

"We know that there are nearly mass-less objects in the universe. One important example is the electron. Obviously, the suggestion of the existence of a nearly mass-less media is a big deal, and will no doubt conjure up the notion of an 'ether.' We will come back to this subject at a later time.

"This Spirographic Grid pattern has profound implications to our understanding of the physical nature of the universe, including the four known forces: gravity, strong nuclear forces, weak nuclear forces, and magnetism.

"Of particular interest is what the discovery of the Spirographic Grid, or we can just simply call it 'Spirogrid,' means to our understanding of the relationship between mass and energy. Later we will discuss a revision to one of the most famous equations known, $E = mc^2$. We will use the knowledge of the Spirogrid to derive a new, more complete, and comprehendible energy equation.

"Also, we will unveil a system of harnessing the energy of the Spirogrid. But first, let's use the understanding of the Spirogrid to address a question that has perplexed humanity for a very long time: What causes gravity?"

GRAVITY

Before he started on the next part of his presentation Manta peered out the window for a second. He could see it was a beautiful day outside. There was a slight westerly breeze and he could see sailboats traversing across the bay in precise and predictable angles relative to one another. He turned his gaze from the bay back to the faces of his group members and began his next point.

"Gravity is the result of what is called a net resultant force. The constantly changing directional motion of the Spirogrid causes a shielding or impedance-type resistance pocket between two masses, which results in this net resultant force. Therefore, gravity is not an 'attraction' but rather a net resultant force pushing two masses toward each other. An illustration showing the mechanism of gravity is provided as Figure 2," Manta said as he scrolled through images of his presentation, and referred to the handout with Figure 2.

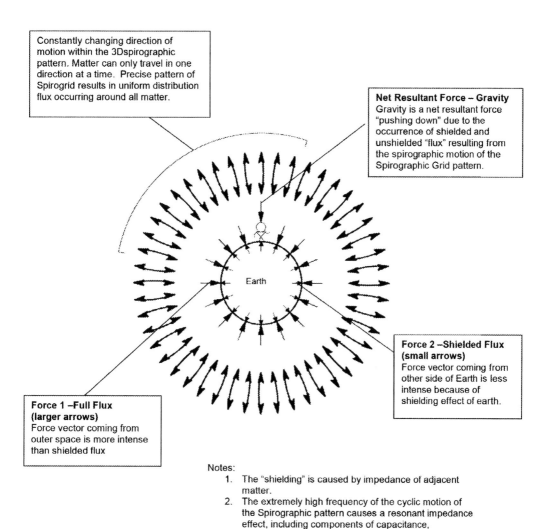

Constantly changing direction of motion within the 3Dspirographic pattern. Matter can only travel in one direction at a time. Precise pattern of Spirogrid results in uniform distribution flux occurring around all matter.

Net Resultant Force – Gravity
Gravity is a net resultant force "pushing down" due to the occurrence of shielded and unshielded "flux" resulting from the spirographic motion of the Spirographic Grid pattern.

Earth

Force 2 –Shielded Flux (small arrows)
Force vector coming from other side of Earth is less intense because of shielding effect of earth.

Force 1 –Full Flux (larger arrows)
Force vector coming from outer space is more intense than shielded flux

Notes:
1. The "shielding" is caused by impedance of adjacent matter.
2. The extremely high frequency of the cyclic motion of the Spirographic pattern causes a resonant impedance effect, including components of capacitance, impedance, and inductance.

Figure 2 – Mechanism of Gravitational Force.
See YouTube video "Artificial Gravity Demonstration".

Manta paused so the members could look over the figure for a few moments to get familiar with it, then he continued.

"This mechanism of gravity is consistent with the observation that the intensity of the gravitational forces diminishes with the inverse-square of the distance. This inverse-squared characteristic of gravity is synonymous with the shielding of light from a light bulb. For example, the intensity, or flux density, of the light from a light bulb hitting you decreases as you move away with the same inverse-squared law as gravity.

"It is easy to illustrate gravity by placing two buoyant balls in a tank of water, where each ball is held half submerged by a string connected to the bottom of the tank, then setting the tank on a vibrating plate operating at a resonant frequency. As the water vibrates, the balls will gravitate toward each other. They move toward each other due to a low-pressure area between the two balls caused by the shielding of the wave energy reflecting in all directions within the vibrating water. The resulting force pushes the two balls together. Since there is a shielding of the media, this explanation of gravity relies on a corporeal motion effect, that is, a medium of some sort, which we will discuss later.

Manta used the multi-media projector to play a video on the screen showing the gravity experiment, where the vibrating plate was turned on and off, showing the two balls "attracting" each other when the plate was vibrating, then immediately coming apart when it was turned off.

"Why don't we take a few minutes to stand up and stretch? Also, please help yourself to a snack or beverage if you'd like."

Manta let the video play over and over while the group talked amongst themselves and enjoyed the quick break to get some coffee and stretch their legs. Manta was organizing his notes for the next part of his presentation.

Once they were all seated, Bill made a gesture to Manta to resume his presentation. He jumped in, "there are other important phenomena occurring due to the spirographic motion of the universe related to the changing center of rotation," said Manta. "For example, we are all familiar with centrifugal force, which is what you feel pulling on the string when you attach a ball to a string and swing it in a circle. The amazing thing about spirographic motion is that, because the center of rotation is always changing, the direction of this centrifugal force is also constantly changing. This changing center of rotation has profound ramifications in the effort to describe the causes of the four major forces that exist in the universe. These ramifications are further addressed in the appendix to the handout I will provide at the end of this meeting.

"One quick example of a centrifugal force caused by a type of spirographic motion is the tidal action of large bodies of water on Earth. Technically, the moon does not rotate around Earth. Both Earth and the moon rotate around a point that is between the center of gravity of Earth and the center of the moon. This point is called the 'barycenter.' This rotation explains why there are two high tides—a higher high tide and a lower high tide—each occurring on the opposite sides of Earth. The higher high tide is caused by the stronger gravitational force between Earth and the moon that pushes the oceans toward the moon. This higher high tide is due to the gravitational force caused by the corporeal motion of Earth relative to a submass media.

"The lower high tide is caused by the weaker centrifugal force of Earth as it rotates about the barycenter. It flings the water outward and away from the barycenter, in the same manner that water would accelerate if you swung a soaked rag around in a circle. Since the centrifugal force causing the lower high tide is due to the rotation of Earth, it is dynamic in nature and does not depend on the corporeal motion of a media."

Manta changed positions and stood next to the podium before continuing.

"Now, after seeing the effects the barycenter has on Earth's tidal actions, imagine the implications of a constantly changing center of rotation occurring due to the spirographic motion on all matter in the universe. This example of centrifugal force is truly a 'force at a distance'—a term used by early theorists in the field of physics and electromagnetism, and will be an important component in discoveries about the causation of forces of nature, including magnetism. Again, more detailed descriptions of how the dynamic and corporeal motion of the Spirogrid are related to the forces of nature, including gravity, are provided in the appendix of the take-home reading assignments."

Manta glanced at the clock to check how much longer he had before they needed to break for lunch. He had a little over an hour.

Manta continued. "All of the gravitational forces and motions that I have described are explained by what is called kinetic energy—the energy associated with a mass traveling with a velocity. It is important to understand that the same kinetic energy that causes gravity also causes violent nuclear explosions. I will try to discuss this with sensitivity, due to the disaster that happened last week.

"Let's start with reviewing how an atomic bomb works. An atomic bomb is made by taking a ball of core material, made of fissile material such as uranium, and surrounding it with an assembly of steel wedges. The steel wedges are made to form a spherical shape surrounding the core material. The spherical assembly is then surrounded with a layer of TNT explosive material. This whole assembly of TNT, steel shells, and fissile material are then encased in a very strong steel shell. The core material, uranium, is the highest molecular weight naturally occurring material on Earth. When the TNT explodes, the symmetrically radial compressive forces of the TNT blast are contained by the surrounding steel shell and are exerted onto the internal steel wedges, which in turn compress the core material to a point called "supercritical." This is what causes the violent nuclear reaction of the uranium.

"The textbooks will tell you that the violent nuclear reaction occurs because there is a chain reaction of the neutrons within the uranium. We've all heard this story. This explanation implies that the energy from the violent nuclear explosion comes from, or is *intrinsic* to, the mass of the uranium. The equation $E = mc^2$ is often cited to illustrate the massive amounts of energy that are *intrinsic* to the mass, specifically claiming that mass is energy and, vice versa, energy is mass.

"Well, the neutron theory and $E = mc^2$ do not accurately explain the atomic explosion. Mass is not energy and the energy from the violent nuclear reaction is not from the intrinsic energy of the mass. Instead, the energy is from the kinetic energy of the mass due to its velocity relative to the Spirogrid. Let me explain," said Manta, after he saw quite a few perplexing looks around the table.

"Let's assume Drew weighs 100 kilograms," Manta said as he tapped Drew Gardner on the shoulder.

"Wait a minute, that sounds a little high to me," Drew said in his Australian parlance. The rest of the group laughed at Drew's predicament. Drew was a big man, and 100 kilos would not be far off.

"Of course, Drew, but let's just go along with it to make the math simple," replied Manta.

Manta moved over to the white board, took a marker, and started writing a calculation as he was speaking.

"We already described how we all have a velocity of 108,000 kilometers per hour as the earth travels around the sun. If we take Drew at 100 kilograms and use the well-known equation for calculating his kinetic energy, that is one-half times the mass times the velocity squared, converting kilometers per hour to meters per second, that comes to," Manta wrote the solution on the board and turned to the group.

"That's forty-five with nine zeros, or, forty-five gigajoules," said Manta. "Now, if we wanted to harness all that kinetic energy, we would convert it to power by bringing Drew's velocity down to zero, and since we are talking about rapid reactions, let's say we do this quickly, in about a tenth of second, which is an eternity compared to the velocity of a thermonuclear explosion. So, divide 45gigajoules by .1 second, and we get 450 gigawatts. To give you an idea how much power this is, the entire United States consumes a total of 80 gigawatts in a whole day! And please realize that this power is only due to the kinetic energy from our velocity as Earth orbits the Sun. Our velocity through the Spirogrid is much much higher, possibly the speed of light.

"This energy calculation is an indisputable fact. And remember, this is Drew's kinetic energy as he sits there doing nothing. So there you go, Drew, I bet you never felt so magnanimous," Manta joked.

"I reckon not," replied Drew.

"The purpose of this simple illustration is to show the incredible amounts of energy due to our dynamic motion through space. I am not saying this dynamic motion is the complete source of energy that is associated with a nuclear explosion. For our meeting tomorrow we will derive a new energy equation that identifies a yet unidentified form of energy in the universe, called 'transformic' energy. And this is a significant form of energy that is associated with nuclear reactions, and all chemical reactions. What I am saying is that to understand the energy associated with mass can be simple high school stuff, and does not require mysterious quantum mathematics, or mass is energy and energy is mass bunch of mumbo jumbo," Manta exclaimed with an air of frustration.

"Our Milky Way Galaxy is flying though space at an incredibly high speed, in the order of the speed of light. It is moving *so* fast that within the microseconds that the mass of the atomic explosion has gone supercritical, our galaxy has changed directions millions of times and has traveled an incredibly far distance. The route that our galaxy has traveled in this split second is not linear, but occurs in a resonating spirographic pattern, you see.

"It is important to understand that the same energy caused by motion—energy related to velocity and mass—which causes gravity is also the energy that causes the violent nuclear explosion.

"Let's go over a nuclear explosion again. When the mass is compressed to a *supercritical density*, it creates a *critical resistance* that impedes the ability of the mass to travel, and change directions, through the Spirogrid. Now remember, this velocity could be dynamic, such as Drew's velocity traveling with the earth around the sun, or corporeal, such as the velocity of the fish relative to the flowing steam of water. This *resistance* to the changing directions causes further compressive forces, the same forces that cause gravity, which cause further resistance, which cause further compressive forces, so on and so forth, resulting in the catastrophic autocatalytic collapsing of the mass into oblivion. This process results in the tremendous release of energy that is observed with a nuclear explosion. The exponential rate of the reaction is due to the autocatalytic collapsing of the mass; in other words, once the collapsing starts, further collapse increases the density of the mass, which further increases the rate of collapsing, and so on, resulting in a blazingly rapid violent release of energy. This is the same mechanism that occurs when a star goes supernovae."

Manta walked over to the computer and pressed a few buttons to bring a visual to the group's attention on the screen in the front of the room. He had a slow motion movie of a nuclear reaction and a star going supernovae with arrows pointing out the phases of what was occurring, which easily explained what he was just talking about.

After a few moments of silence, Malik spoke up before Manta could continue again. "Miles, are you saying that Einstein was *wrong*?!" asked Malik, not able to hide his shock.

Manta walked back to behind the podium in front of the table and slowly answered, "Yes. I am saying that Einstein was wrong."

Manta noticed not only Malik's surprise to his answer, but he saw other members exchange glances with each other with very wide eyes.

Manta continued before their thoughts raced too far. "There is no doubt that Einstein was a great theoretical physicist. He himself felt that he struggled and searched, and he himself believed he did not have all the answers. It was the circumstances of that period of time that caused modern scientists to fall in love and make him into the infallible god of physics he is known to be.

"Remember, Einstein did not invent nuclear power. That credit mostly goes to the work of Enrico Fermi, and then later Robert Oppenheimer.

"The high-level physics business that we know today was brought about by men like Edward Teller, the so-called father of the hydrogen bomb. If you really look at what men like this were about, the word humility would not come to mind. Science took a wrong turn during the period of the Cold War. The 'secrets' of the Cold War led to more deception and scams than truth about core science. The Cold War brought about a lot of manipulations by the defense industry to sell governments their wares and schemes, such as the neutron bomb. Of course, these manipulations were justified if they caused the Russians to be deceived, or go broke trying to counteract the threat. Oh, and by the way, according to the Spirogrid theory, there is no such thing as the hydrogen bomb or fusion energy. At least not the way conventional science describes it."

"What?! How can you tell us that there is no H-Bomb?" cried Takashi from the back of the table, clearly feeling aghast.

Manta walked around to the side of the podium and tried to explain this rationally. "We know that the 'fusion' hydrogen bomb uses a fission bomb to supposedly ignite the thermonuclear reaction, in the same way that the TNT starts the fission bomb. Tell me, Takashi, who is going to dispute the claim that the nuclear bombs that they set off in the Bikini Islands were fusion devices rather than plain old fission bombs? Name one person that is going to testify in front of Congress saying that Edward Teller was wrong. Name one person who can obtain the data and is willing to throw away all of their life's work and stalwart belief in fusion, to say fusion does not exist."

Manta made sure he didn't raise his voice and he returned to behind the lectern and continued.

"Remember, these are boys with toys. Try to take away a child's favorite toy, and you will understand what I mean. Disclosure that fusion energy does not exist is a direct threat to their work and livelihoods as scientists, and all their aspirations to unlock the energy that powers the Sun. This scenario has repeated itself over and over throughout the history of humanity, including even our most recent history. In the 1970s the public was duped by the famous 'Glomar Explorer' built by Howard Hughes, but funded by the US Government, to supposedly mine the bottom of the ocean, where it was later revealed it was built to retrieve a sunken Soviet nuclear submarine. In the minds of the controlling powers all these lies and manipulations are justified. "

"Then what are those giant explosions we see? Isn't that proof enough that there are hydrogen fusion bombs and not Nagasaki-style fission bombs?" asked Drew from the middle of the group.

"There are many shapes, forms, and sizes of bombs," Manta began to explain. "This is also true of conventional explosives. The environment where the bomb goes off is especially important. For example, those magnificent blasts that occurred in the Bikini Islands…yes, they looked different from the land-based bombs. Not because they functioned by fusion, but because they were set off underwater. Also, they were much larger devices. Again, the H-bombs are nothing more than large fission bombs. The mechanism of how it works is a critical disruption and resistance to motion in the Spirogrid, and the energy released is not intrinsic to the mass. Rather, the energy released is from the original velocity of the mass, including translational and rotational, and as we will derive later, a form of energy called transformic energy."

Drew felt that answered his question sufficiently enough and relaxed back into his chair once again.

"The other example of fusion would be the attempts to build fusion reactors," Manta explained. "When I was a teenager, and even into my college years, there were claims that, in the near future, fusion energy reactors would make electricity so cheap that homes would not even require a meter to determine electric usage and fees. The scientific community kept promising ten more years, ten more years, ten years from now…Yet, here we are. Despite billions of dollars spent, there is not one demonstrable example of electricity from fusion power. Not even close. Instead, we are scrambling to secure more sources of natural gas to power our grid. Clearly, we should have put that money into solar cells and biomass."

Manta saw Katerina shake her head agreeably to what he had just said and he heard a lot of "hmms" around the table as well, as they all sat back in their chairs and continued to listen.

"I am sorry to be the bearer of bad news," continued Manta. "But it really does not have to be considered bad news…

"I am afraid that I painted the wrong picture about scientists and engineers by exposing a few cases where we took a wrong path. By far, in most cases, scientists and engineers are sincere about their efforts to improve the world through innovation and discovery. It is very important for everyone to understand that this is not the time to lose faith in science.

"Yes, this information should all be taken as good news. I have not even begun to describe the implications of the Spirogrid to other more applicable areas of science. The greatest task at hand is going to be figuring out how to disseminate the information in a responsible and effective way."

"Miles, aren't you afraid that, by revealing this information, you are going to educate every terrorist group in the world that is trying to build nuclear weapons to only be able to build a better bomb?" asked Walter, who felt very concerned.

"Well, I keep going back to the point that the biggest barrier to preventing people from building atomic bombs is the simplicity," said Manta. "The truth is that if any rich nations want to build a bomb, they will eventually be able to do it. The main issue faced by any group or nation that wants to build a bomb is that, if they use it, they are signing their own death warrant. Any use of a nuclear bomb will result in a severe retaliation and probably a nuclear annihilation. The capability that we have to worry about is associated with the means of delivery. And this power is in the hands of only a few nations. More effort needs to be made on mutually beneficial methods of deterring nations or groups from wanting to use these weapons.

"Rather than hoarding technological advancements, the world needs to move forward in advancing the living conditions of all people. When that happens, fewer desperate people would cause problems for their neighbors or the environment. A greater respect for Earth needs to be shared by all."

Bill saw a pause he could use in Manta's speech to recommend that the group should take break. He looked at the clock, then announced, "Let's take a break for lunch, if this is a good time for you, Miles."

Manta agreed and felt hungry himself.

"OK, let's meet back here in about an hour. Down the hall is a fully catered luncheon, compliments of Mr. Manta," said Bill.

The group members left the room to grab lunch and get some fresh air outside on the convention center's patio. Miles made a plate, went back into the conference room, grabbed a chair, and moved it over near the window so he could gaze out at the grand city while he ate.

Bill was the first person to come back from the lunch break, and he came over and stood next to Manta by the window.

"How am I doing?" Manta asked him with a half smile.

"Too dramatic, but we all know how passionate you can be," said Bill, smiling back.

Eventually all the members of Group T9 had reentered the room and found their places around the table again. They felt like they were on information-overload, but were completely enthralled with all that Manta was sharing. They were ready for Manta to continue.

ORDER IN THE UNIVERSE

Manta was walking back from throwing out his trash in the can in the back of the room while everyone was taking their seats again. He walked right to the podium and jumped back in where he had left off.

"Before I talk about other scientific implications of the Spirogrid, I would like to touch on some philosophical dilemmas that we are going to be facing with the revelation of this information," he said.

"As you are aware, the general theory about the origin of the universe revolves around the Big Bang Theory. With our technological breakthrough, the energy associated with the origin of the universe will be part of a New Energy Equation that we will derive from our knowledge of the Spirogrid. Regarding the Big Bang Theory, to be honest, I don't know if that theory is valid or not. But I am certain that it does not affect my life, or how I or most people in the world put bread on the table.

"There is one flaw that I see with the Big Bang Theory: it implies that there is a disorder, or even chaos, in the universe.

"I am very puzzled that scientists study the objects of nature that are visible to the naked eye and also the microscopic world with electron microscopes, and they see incredible order. Yet, when they look up to the heavens they assume—randomness or chaos. Why?" Manta asked and looked around the room for a brief second.

"When you look at the simplest organisms—say, a fruit fly that seemingly comes from nowhere—the closer you look, the more you see fascinating complexity." He explained, "Each eye segment or each hair follicle looks more and more detailed. The higher the magnification, the higher the order of complexity. Even the subatomic world shows amazingly intricate crystal lattice structures. Yet, look up at the universe, and many assume it is—random chaos. Why?" Manta asked again.

"Is the vast universe governed by disorder?" he continued. "Not at all. In fact, the incredible order and complexity of the motions and orbits in the universe are the most stunning examples of order we can observe. The order and precision of the 3D spirographic pattern of the Spirogrid are mindboggling. Take a look at Figure 1 again and see what I'm talking about.

"The mathematical challenges of modeling the intricate motions of the Spirogrid are sobering. Yet, there are brilliant mathematicians that produce brilliant models and simplify the problems—much like Kepler and Newton simplified the elliptical planetary orbits, or how Tesla modeled alternating current electricity.

"I do not care if you believe the formation of the universe was an accident, by chance, or the work of an Intelligent Creator. The order and complexity of the universe is undeniable. And it is a wondrous thing.

"At this point, I'd like to show you a brief movie clip highlighting some of the order and complexity I have been talking about. I know some of you are visual learners, so this will help illustrate what I have been referring to."

Manta walked over to the computer, clicked a few buttons, dimmed the lights, and the movie began on the screen.

The group members enjoyed the artistic short film that highlighted the universe's order from the widest view of the galaxies down to the microscopic details found under the most advanced scanning electron microscopes in the world. Manta turned the lights back up and began where he left of.

"As I mentioned before, the understanding of the Spirogrid forms the basic explanation of the state of the universe and the all other scientific laws, phenomena, explanations, and observations," he stated.

"In order for you all to be effective in administering this information, I believe that you must have a rudimentary understanding of some of the most important discoveries related to the Spirogrid. Heaven forbid that you find yourselves depending on someone else to explain these things to you. And I believe you will find the underlying science of these discoveries to be easy to understand."

Many members looked at each other and rolled their eyes.

Manta continued. "The best approach is to use a historical perspective. I know that most of you are not scientists, so I have prepared this handout titled 'Handout A—Newton's Second Law and $E = mc^2$ Revisited' for those of you that have a background in calculus," said Manta as he passed around the packet to each member, whether or not they had a mathematical background or not.

"Please read this before the next meeting," he said, making his way around the table. "For those of you who do not have a background in calculus, just glance over it and be prepared to see some basic algebra during the next meeting.

"Also, for the next thirty minutes, I would like to show a video that describes and illustrates the Spirogrid and its relationship to the scientific laws. I think you will glean much understanding through this presentation."

Manta clicked a few buttons on the computer and dimmed the lights again before he started the video. The group members angled their chairs toward the television screen and sat back to appreciate the opportunity to be informed about this subject matter. They were all very intrigued.

Once the video finally ended, the meeting was about to be over. Manta facilitated a brief discussion about the topics presented in the film and clarified any confusion about the Spirogrid.

It was now time to see what the next move would be. Bill took the floor and had the group vote whether or not they would continue the series of Manta's presentations.

The group voted unanimously in favor of continuing the series, and they scheduled the next meeting to occur the very next day, Tuesday. They packed up their belongings and slowly trickled out of the conference room. Manta let Bill know he was taking the flash-drive recording of the meeting, and they walked out together.

CHAPTER 7

E ≠ MC² – A NEW ENERGY EQUATION

Everything was unfolding just as Manta had predicted. Still, no person or group had claimed responsibility for the atomic blast in Chebala.

On a positive note, the incident was creating profound changes in the relationships between many countries. There was a high level of cooperation between all the nations.

It has been said that calamities changed the minds of men. Or, was it just plain fear? There was a global consensus that this must not be allowed to happen again.

TREK MEETING NUMBER TWO

It was Tuesday morning. Manta arrived at the conference room earlier than the group so he could write a few things on the board and pin up a figure of the Spirogrid.

The members started to trickle in slowly, and they enjoyed a few morning snacks and a cup of a coffee before they took their seats. Bill, who acted as the chairman protem, sat in the front right hand seat again and brought the T9 meeting to order.

"I don't know about the rest of you, but I certainly had a lot of fun with the reading assignment last night," Bill said facetiously.

"I thought you would like it," said Manta. "Nothing like a little calculus drill to cozy up with next to the fireplace?"

"Yes, definitely brilliant," said Bill.

The others nodded their heads.

"It was like cramming four years of college into four hours of hell," teased Bill, laughing. "But I got it OK. I think I passed."

A few of the other members felt the same way and chuckled.

"I hope all of you get to a comfort level with the concepts," Manta explained. "Remember, I am not asking you to become mathematicians. Mostly, I wanted you to know the history and purpose of mathematical calculus. Please keep in mind, you will have plenty of time after our meetings to conduct internet searches and review the text books for better explanations describing the historical significance of an important mathematical milestone, namely, the use of infinite and infinitesimal limits to solve problems related to the area-under-the-curve and instantaneous-velocity controversy."

"Miles has explained to me that we are going to have a long lecture today, so we have food and drinks at the back of the room," explained Bill. "So don't hesitate to help yourselves at any time. Any one of us getting up and moving to the back of the room will automatically come with a short break that will allow all of us to stretch or use the bathroom," explained Bill.

"OK, Miles, you have the floor. Let's get this party started," said Bill.

"Thanks, Bill," Manta said and he began.

"When I first spoke to you about my revelations yesterday, one of your first questions was, 'Are you saying Einstein was wrong?'

"Putting yourself in my shoes, and knowing what I have to disclose, shouldn't this comment be expected? Any work that presents new revelations about the nature of the universe will immediately be compared to the work of Albert Einstein.

"Therefore, you can understand why it is important for our education about the Spirogrid," Manta walked to the board behind him to point to the figure of the grid he had posted, "to start off with an evaluation and comparison to one of the most famous mathematical equations of our modern era, $E = mc^2$. In order to do this, we have to achieve some degree of understanding of the tools that brought this equation to us.

"For those of you with a background in calculus, the handout I gave you yesterday provides a sample of how $E = mc^2$ could be obtained from the derivation process of calculus applied to Newton's second law."

The group was intrigued from the start. Especially since Manta had "$E \neq mc^2$" written on the board. They sat up attentively and some had the previous day's handout in front of them for reference.

A NEW ENERGY EQUATION (AREA UNDER THE CURVE)

Manta began the next portion of his presentation. "For those of you that do not have a background in calculus, I am happy to let you know that there is an easier approach," he said. "Amazingly, this easier approach provides a much broader mathematical expression than $E = mc^2$. You will see how the New Energy Equation is much easier to understand and how it accounts for all of the energy in the universe.

"To accommodate the discussion of this material, I will use a historical perspective.

"I am going to describe things using terms that you may not be familiar with, but should be comprehendible by a high school level student, so don't worry. Just bear with me, and with a little time and help, you will get the gist of the main points."

****

Manta's Lecture

This lecture is supposed to be entertaining. Whether you have a background in mathematics or not, it is recommended that you first read through this material from beginning to end in a cursory manner, possibly making notes, then go back later for a more detailed study.

****

AREA UNDER THE CURVE, PROBLEM AND SOLUTION

Before we go into the mathematical exercise of deriving an energy equation from the discovery of the Spirogrid, we must acknowledge that we have the privilege of standing on the shoulders of giants. In a sense, this means we get to cheat.

Unlike the scientists from hundreds of years ago, we can do things without having to prove exotic equations or overcome algebraic dilemmas that threaten to stop us in our tracks. We have the benefit of previous work to provide us the tools to solve mathematical problems.

However, in order to produce our New Energy Equation, we do need to understand what those tools are. One tool that took scientists about one hundred years to figure out is related to the "area under a curve."

Let's start by describing something called a linear equation. Don't worry; it is simple. In fact, you use it every day. For example, if you were buying 2 gallons of milk at $4 each, you know that you would be paying $8 at the cash register. And, if you were buying 3 gallons of milk, you know you would be paying $12. That's all there is to it. You fully understand and have mastered a linear equation. Again, you do this every day with various objects and quantities and you may not even realize that you are using linear equations.

Another example of a linear equation is related to speed, or velocity. Say you are driving at 100 kilometers per hour. How far would you go in 2 hours? Yes, 100km/hr x 2 hrs = 200 km.

This is called a distance and rate equation, and scientists and engineers always keep it in the back of their minds as d = r x t, or, more often, d = vt, where d is distance, v is velocity, and t is time.

You might think that that the Romans and Egyptians would have known and understood this, but we tend to take simple mathematical expressions for granted. The concept of velocity in terms of a mathematical expression of distance divided by a unit of time would have been foreign to them. They would have understood distance in terms of the number of a day's journey, but probably not in terms of velocity multiplied by time.

However, let's move ahead in time to the days of Galileo and Newton, when cannon balls were being fired at great velocities and the planets were understood to be rotating around each other. Back then, people were dealing with objects traveling at much higher velocities. The rate equation would have become more widely used. And with that came the discovery of a major limitation to its accuracy.

Notice that the use of the linear equation d=vt, in the example of you driving your car for 2 hours, only accounts for the distance you are traveling at the constant velocity of 100km/hr for 2 hours. Always, and in every case, when you take a trip in your car, you start off at zero velocity. Therefore, the d = vt formula only works between two points on your trip—when you are "up to speed," and where your velocity is constant.

What about the beginning of your trip, where you were speeding up, or, as scientists say, accelerating? How do you determine this distance? It may seem simple, but, as we will see, it gets complicated.

If you were a serious thinker, like Galileo, and you knew the time it took your car to reach 100km/hr, you could approximate the distance covered while accelerating by dividing the time portion into little equal segments and multiplying that little time segment by the average velocity for that time segment between 0 and 100km/hr. Stay with me, and in a minute we will use some visual aids to help you understand what I am saying.

For example, when you started on your trip, let's say you knew it took 10 seconds to reach 100km/hr. Therefore, you could assume that in the first second you were traveling an average of 10km/hr, the second second you were traveling 20km/hr, the third second you were traveling 30km/hr, and so on. The distance you traveled during acceleration is the summation of the distance traveled in each of these 10 segments.

So, doing the math, it would look something like this:

Distance traveled during acceleration = the summation of [v • Δt] between 0 and 100 km/hr

= Σ[10km/hr • Δt] + [20km/hr • Δt] + [30km/hr • Δt] + [40km/hr • Δt] + [50km/hr

• Δt] + [60km/hr • Δt] + [70km/hr • Δt] + [80km/hr • Δt] + [90km/hr • Δt] + [100
km/hr • Δt]   Where Δt = 1 second.

Note: The symbol " $\sum$ " means "the summation of" in the language of mathematics.

This was the kind of approach Galileo, and others, where doing to solve the problem of distance traveled during acceleration. And, to analyze their approach, we will make a table showing the data, and plot what the velocity versus time relationship looks like on a graph, as shown in Figures 3A and 3B.

Note: In order to multiply by the Δt = 1 second for Table 1, we will convert the velocity from km/hr to meters per second.

Elapsed Time (seconds)	Velocity at Interval (km/hr)	Velocity at Interval (meters/sec)	Distance at Interval (meters) d = velocity x Δt Where Δt = 1 sec	Cumulative Distance at Interval (meters)
1 sec	10km/hr =	2.78 m/sec	2.78 meters	2.78 meters
2	20	5.56	5.56	8.34
3	30	8.34	8.34	16.7
4	40	11.1	11.1	27.8
5	50	13.9	13.9	41.7
6	60	16.7	16.7	58.4
7	70	19.4	19.4	77.8
8	80	22.2	22.2	100.0
9	90	25.0	25.0	125.0
10	100	27.8	27.8	152.8

Figure 3A-Table of calculated data for 10 seconds of acceleration to velocity of 100 km/hr (27.8 meters/sec).

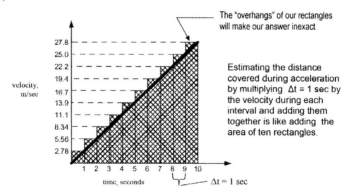

The "overhangs" of our rectangles will make our answer inexact

Estimating the distance covered during acceleration by multiplying Δt = 1 sec by the velocity during each interval and adding them together is like adding the area of ten rectangles.

Figure 3B - Graph of velocity vs time for 10 sec of acceleration to velocity of 100 km/hr (27.8 meters/sec). Note: We could use the area formula for a triangle (½bh), but our simple example is meant to illustrate the approach that led to the discovery of calculus, which enabled the analysis of non-linear curves.

According to this method, the cumulative distance traveled after ten seconds of acceleration is 0.1528 kilometer (152.8 meters), see cumulative distance column in Figure 3A.

But now we have another problem. Did you notice that our answer is an estimate and not an exact answer to the problem? It is an estimate because we used the average value of the velocity that occurred in the 1 second subinterval (actually, it is the value of the right-hand endpoint of the subinterval, but we want to keep our illustration as simple as possible). Using a little intuition, you can see our error graphically by the little areas overhanging the line in Figure 3B, which would add up to make our distance answer higher than the true value.

It should be obvious that we could make our answer more exact if we decreased our Δt by half, or even less. In fact, it is true that the smaller the Δt, the greater the accuracy of our answer. Ultimately, the most accurate answer would be where Δt is infinitesimally small, or even zero!

But now we have created two potential problems. If Δt is infinitesimally small, we have to make an infinite number of additions. And that's not possible.

And the second problem is even worse. If we make Δt equal to 0, the whole calculation falls apart, because multiplying any number by 0 will equal 0, and we know that we did not travel 0 distance during the 10 seconds. This whole concept of instantaneous time was a perplexing problem for decades. How can you have a value for instantaneous velocity, with units of distance divided by time, when instantaneous means time = 0, putting a 0 value in the denominator, when you cannot divide by 0? It is against the rules of algebra, and a seemingly complete dead-end.

Could we ever get an exact answer? Maybe it's not possible.

Well, it is possible! And, as mentioned before, it took a hundred years to figure it out. The credit goes to two men: Isaac Newton and Gottfried Leibniz. They solved the problems by introducing the concepts of limits, derivatives, and integrals. Essentially, their trick worked because the use of limits allowed the cancelling out of the 0 in the denominator of velocity as the limit approached 0 (a trick of dividing a limit by itself to make it equal to 1, or, in the language of mathematics, making it cancel out and go away).

So did you notice that when you look at our analysis graphically, which is shown in Figure 3B, the value of the distance is equal to the area under the curve made by plotting velocity verses time? Or, does it? We implied this earlier when we intuitively analyzed our estimate of the distance covered during acceleration and we referenced that the addition of the overhangs would make our distance value higher.

OK, we cheated a little on this. We never had the right to assume that the area was equivalent to distance. Hundreds of years ago you could not just make this assumption, you had to prove it. The use of the term *area* brings to mind units of area, such as square meters of floor tile—not distance. Try to put yourself into the minds of people five hundred years ago and just think of the overhangs as obviously adding quantity to the answer and not necessarily distance.

Early mathematicians could not assume that the area under the curve equaled the exact distance. They probably argued this until they were blue in the face. Until calculus was invented, they could not even calculate the exact area under the curves generated by plotting velocity versus time, for the reasons described earlier about the Δt equal to 0. You have to understand that a lot of this stuff is not intuitive to everyone, and it is subject to high levels of scrutiny. For example, a Seventeenth-Century contemporary of Newton and Leibniz, George Berkeley, the Bishop of Cloyne, called the limits "ghosts of departed quantities" and dismissed them as nonsense.

The calculus proved to be true, correct, and exact. Yes, the area under the curve *is* the exact distance, in both value and units of measure, and the value of the integral, a mathematical term that we will talk more about latter, represents the exact area under the curve. Amazing, but true: it took over a hundred years of effort for this simple understanding to be confirmed. So, don't feel too bad if you don't understand all of this material at first. You can take your time and review the handouts and conduct internet searches to get all the background information.

There is one more observation we need to make about our velocity versus time versus distance analysis. Please look at the data table of Figure 3A and notice that, although the velocity is increasing linearly with time (constant acceleration), the distance is increasing exponentially with time. A curve showing the exponential relationship of distance versus time from our data in Figure 3A is provided as Figure 4.

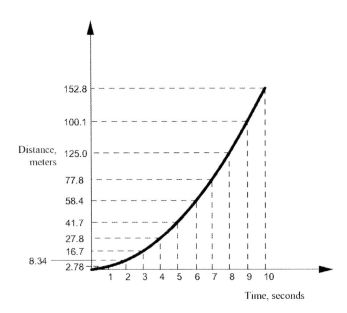

Figure 4 - Graph of distance vs time for 10 sec of acceleration to velocity of 100 km/hr (27.8 meters/sec). Distance is increasing exponentially, as a function of t². The process of calculus was invented to analyze this type of relationship, but other surprises were discovered along the way, including the awe-inspiring and nearly magical "constant of integration".

Unlike linear relationships, exponential relationships are very difficult to under-stand intuitively. Typically, you can't just rattle off an exponential relationship from your head. For example, it's not as easy as when you were buying 2 gallons of milk at $4 each and you knew you would be paying $8 total.

Figures 3A and 3B show a linear increase in speed of the car over the 10-second time period. This is expected because we defined the velocity to increase by a con-stant rate of 10 km/hour per second, which is equivalent to a constant acceleration of 2.78 m/s² (this is determined by dividing 10 km/hour by 1 second, keeping in mind that you have to convert 10 km/hour into units of meter/second).

Table 3B and the graph of Figure 4 show how the accumulated distance is in-creasing by a factor of t². If you have some background in algebra, you can see that the distance versus time graph in Figure 4 is a parabola-shaped curve depicting an exponential relationship.

Galileo and others observed this nonlinear relationship of the distance versus time in their experiments of rolling a ball down a ramp and logging the data. It was a very perplexing problem for them to analyze. In a similar and parallel story to the "area under the curve" problem, it wasn't until the mathematical tools of Newton's and Leibniz's calculus that we had formulas to determine the exact values for these exponential relationships.

Fortunately for us here today, we do not need to go too deep into these more complex exponential relationships. The reason I am introducing these more complex mathematical relationships is to make the point that it was the invention of calculus that allowed us to understand, and quantify, what was happening to objects that were in the state of acceleration. In the journey of developing these new mathematical methods, unexpected discoveries were made, including the awe-inspiring and nearly magical "constant of integration".

To accelerate our discussion, Figure 5 provides the calculations that allow us to determine the exact area under the curve, and, more specifically, how this is done for the distance equation, an equation that is used by every first year physics student. The derivation of these formulas uses the higher form of mathematics learned in calculus, but can be done with the same confidence as one would have with the use of addition, subtraction, multiplication, and division. For those of you that need a basic introduction, or a refresher, of the mathematical concepts used in our derivation, please review the Precursor to Figure 5.

Precursor to Figure 5 This table provides a review of symbols, operators, and definitions used in mathematics	
+,-, x, ÷	Addition, subtraction, multiplication, and division. These are basic functions that people use every day.
$\dfrac{x}{x} = 1$	Math Law. Any number (x) divided by itself is equal to one.
$\dfrac{0}{x} = 0$	Math Law. Zero divided by any number is equal to zero
$\dfrac{x}{0}$ is undfined	Math Law. Dividing by zero is not allowed, it is against the rules of mathematics. This "menacing zero" creates a conundrum with the notion of "instantaneous velocity" because velocity is distance divided by time, and "instantaneous" means the value of time is zero. The search for an answer to this problem led to the discovery of calculus. Involved hundreds of years of effort to solve the problem.
$\sum f(x)$	The symbol used to express "summation" of a bunch of calculations of a function f(x). The function in our accelerating car example was distance equals velocity multiplied by time, d = vt. Ok for approximation, but not exact due to the "overhangs".
$x \times 0 = 0$	Math Law. Any number multiplied by zero equals zero. This creates a conundrum in our example of determining distance of a car accelerating for 10 seconds because using "summation" we cannot multiply by zero to calculate the exact area under a complex curve.
$\displaystyle\int_1^2 f(x)$	The symbol for "integration", a powerful tool of calculus, similar to "summation". Miraculously, and simply, calculates the exact area under the curve of f(x). Discovered by an algebraic trick where the "limit" of the "menacing zero" is divided by itself, like x/x=1, and cancels out. Also reveals the miraculous "Constant of Integration" where unknown quantities are revealed, as we shall see in Figure 5, including all energy in the universe as shown in Figures 6, 8, 9 and 10.
Force F $= \dfrac{dmv}{dt}$	Process of "Derivation". Not part of Figure 5, but described in Handout A, where "Product Rule" of calculus requires the following "next step" if both terms are not constant: $$F = m\dfrac{dv}{dt} + v\dfrac{dm}{dt} \text{ , } \quad \text{or,} \quad F = ma + v\dfrac{dm}{dt} \text{ , }$$ where "ma", mass x acceleration, is the most famous mathematical relationship of all time, and dm/dt, or "change in mass per unit time", and constant "v", are tantalizing factors for some to challenge the concept of "conservation of matter", and mass-energy equivalence.

Figure 5 - Distance Formula (Distance, Velocity, and Acceleration)

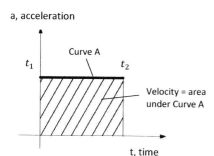

a, acceleration

With reference to Curve A (actually a line in this case), using the modern expression for Integration, we can write:

$$\text{Velocity} = \int_{t1}^{t2} a \, dt = \textbf{Area Under Curve A}$$

This equation says: "the summation of an infinite number of tiny rectangles between the value of t_1 and t_2 equals the exact area under the curve, which equals the velocity of the object".

The result of this integration is:

$$\textbf{Velocity = v = at} + v_0$$

Where, v = velocity after acceleration, a= acceleration, t=time, and v_0= velocity before acceleration, which is the "constant of integration" that must automatically be added, see Note A.

Note A: The process of integration requires that a constant, v_0, be added, which automatically accounts for the possibility that the object may have already been moving at t_1. Like magic, the integration process knows that the **total** velocity of the object is equal to the summation of the velocity before and after the acceleration that occurred between t_1 and t_2 .

v, velocity

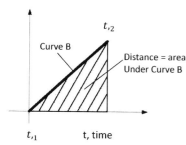

With reference to Curve B, using the modern expression for Integration, we can write:

$$\text{Distance} = \int_{t1}^{t2} v \, dt = \textbf{Area Under Curve B}$$

From above, we have,

$v = at + v_0$, substituting into equation, we have

$$\textbf{Distance} = \int_{t1}^{t2} (at + v_0) \, dt$$

This equation says: "the summation of an infinite number of tiny rectangles between the value of t_1 and t_2 of the equation $(at + v_0)$ equals the exact area under Curve B, which equals the exact distance the object has traveled".

The result of this integration is:

$$\textbf{Distance = d} = \tfrac{1}{2} at^2 + v_0 t + d_0$$

Where, d = final distance, a= acceleration, and v_0= velocity before acceleration, and d_0 = distance form origin before acceleration and v_0 occurred, which again is the constant of integration that must automatically be added, see Note B.

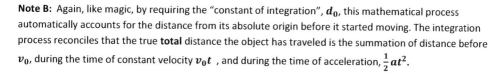

Note B: Again, like magic, by requiring the "constant of integration", d_0, this mathematical process automatically accounts for the distance from its absolute origin before it started moving. The integration process reconciles that the true **total** distance the object has traveled is the summation of distance before v_0, during the time of constant velocity $v_0 t$, and during the time of acceleration, $\tfrac{1}{2} at^2$.

Therefore, using the distance equation provided in Figure 5, we can find the exact answer to the question: How far would you travel if you drove for 2 hours at 100 km/hr, starting from a dead stop, and it took you 10 seconds to reach 100 km/hr?

Distance Equation: $\qquad$ Distance = d = $\frac{1}{2}at^2 + v_0t + d_0$

a = acceleration = .278 m/s² (10 km/hour increase per second over 10 seconds)

v_0 = velocity before acceleration = 0 (started from dead stop)

d_j = distance from origin before acceleration = 0 (our zero distance reference is the starting point at dead stop)

d = $\frac{1}{2}(2.78)(10)^2 + (0)(10) + 0$ = 139 meters = 0.139 kilometers during acceleration

(Exact area under the curve)

Total Distance = distance during acceleration + distance during constant velocity

= 0.139 m + (100 km/hr)(2 hrs) = 0.139 m + 200 km = **200.139 km (Exactly)**

THE MYSTERIOUS "CONSTANT" OF INTEGRATION

Sometimes when you play with mathematical equations strange things happen that seem to defy explanation. This would apply to the constant of integration, which are the v_0 and d_0 values in Figure 5. I'm sure the constant of integration can be described by some complex explanation based on the fundamental rule of calculus that states, "The integral of the derivative is the function of...blah, blah, blah."

But the fact remains that the integration process automatically, and magically, adds in a quantity that we would not intuitively know to exist. This is impressive enough when you are describing the physical placement of an object, such as with a distance and velocity problem. But it is even more impressive when the constant of integration describes a more abstract quantity, such as higher orders of energy, or, possibly, the original state of all energy in the universe, which we shall see in our derivation of a New Energy Equation.

The constant v_0 is automatically added when we integrate $\int a\ dt$ because the integration process automatically accounts for the possibility that the object that is being accelerated may already be moving. For example, say we were riding on the deck of an aircraft carrier that is traveling at 10 km/hr (2.78 m/s) and we could precisely throw a ball forward with an acceleration of 5 m/s² for 1 second. The velocity of the ball due to throwing is 5 m/s, which is the acceleration multiplied by time (a·t) value. But the integration automatically adds the v_0 (2.78 m/s) that we were already traveling at with the speed of the aircraft carrier. Therefore, the true velocity of the ball relative to the ocean, is 5 m/s + 2.78 m/s = 7.78 m/s (28 km/hr).

The constant d_0 is even more impressive. It is automatically added when we integrate $\int at + v_0$ because the integration process not only automatically assumes we were moving when the acceleration started, but it also assumes we were not in our original position. For example, let's say when we threw the ball while we were on the deck of the aircraft carrier, we were 10 kilometers (10,000 meters) away from home, which was our origin. Therefore, after that 1-second period of acceleration, our new position is as follows:

$$d = \tfrac{1}{2}at^2 + v_0t + d_0 = \tfrac{1}{2}(5)(1)^2 + (2.78)(1) + 10,000$$

= 10,005.28 meters (10.00528 kilometers) from the origin, exactly.

Again, this may not seem impressive for a distance problem, but, as we shall see in our derivation of a New Energy Equation, when the constant of integration describes more abstract quantities involving higher orders of energy, and possibly the original state of all energy in the universe, it can provide astounding revelations.

INERTIA VERSUS MOMENTUM

In our final step to derive our New Energy Equation, we will use our "area under the curve" knowledge from the distance equation and apply it to Newton's discovery of a quantity called *momentum*.

During Newton's time, there was a lot of discussion about inertia, but the concept never got very far.

To this day, inertia is a unit-less concept. We do use a term called "moment of inertia," with the units of kg-meter², but this describes the geometric relationship of mass around a point, not a form of energy. We will discuss more about the significance of the moment of inertia later.

Instead of using inertia, Newton used momentum to describe the "energy" of a moving mass, and defined it simply as the mass times the velocity:

Momentum (or M) = Mass x Velocity

This may seem simple enough. But remember, to obtain momentum, Newton had to first define mass. He did this when he discovered the relationship between mass, gravity, and weight (force). This was part of his second law of the *Principia*, which was no doubt the greatest discovery in the history of science. (The first law of the Principia was about how a body in motion stays in motion unless acted upon by an external force; and his third law was about how for every action there is an equal and opposite reaction.)

The mathematical relationship from the second law (the change of momentum, with respect to time, is equal to force), is what we show in the handout, showing how it can be used to formulate $E = mc^2$ with the mathematical process of "derivation".

So, we now know where the word "momentum" comes from, and what it means.

We are now going to use the same process we used to derive our distance formulas in Figure 5 to calculate the area under the curve for the momentum versus velocity curve. And guess what? The area under the momentum versus velocity curve is equal to Energy. We are now going to stand on the shoulders of giants and borrow the mathematical process of "integration" to formulate a New Energy Equation.

Figure 6 - The New Energy Equation - Derived from exact procedure of Figure 5.

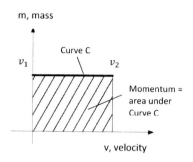

m, mass

Curve C

v_1 v_2

Momentum = area under Curve C

v, velocity

With reference to Curve C (actually a line in this case), using the modern expression for Integration, we can write:

Momentum = $\int_{v1}^{v2} m\,dv$ = Area Under Curve C

This equation says: "the summation of an infinite number of tiny rectangles between the value of t_1 and t_2 equals the exact area under the curve, which equals the momentum of the object".

The result of this integration is:

Momentum = M = $mv + M_0$

Where, m = mass, v = velocity, and M_0= momentum before the objects momentum was changed, which is the constant of integration that must automatically be added, see Note C.

Note C: The process of integration requires that a constant, M_0, be added, which automatically accounts for the possibility that the mass may have already been moving before v_1 occurred. Like magic, the integration process knows that the **total** momentum of the mass is equal to the summation of the momentum before and after the change in velocity between v_1 and v_2 .

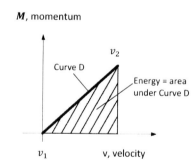

M, momentum

v_2

Curve D

Energy = area under Curve D

v_1 v, velocity

With reference to Curve D, using the modern expression for Integration, we can write:

Energy = $\int_{v1}^{v2} M\,dv$ = Area Under Curve D

From above, we have,

M = mv + M_0, substituting into equation,

Energy = $\int_{v1}^{v2}(mv + M_0)\,dv$

This equation says: "the summation of an infinite number of tiny rectangles between the value of v_1 and v_2 of the equation $(mv + M_0)$ equals the exact area under Curve D, which equals the exact Energy of the mass".

The result of this integration is:

Total Translational Energy = E_t = $\frac{1}{2}mv^2 + M_0 v + e_0$

Where, E_t = total translational energy of mass following a "line or curve", m= mass, and v= present translational velocity of mass, M_0 = momentum of mass before added velocity, and, e_0 = added constant of integration, energy of mass at origin. See Note D.

Note D: Again, like magic, the "constant of integration" automatically accounts for the absolute origin of the energy, e_0, before the effects of added velocity. In light of the discovery of the Spirographic Grid, this new energy equation has profound implications for our understanding of mass and energy in the Spiroverse. A further analysis shows that this new equation for total translational energy, E_t ,is the summation of Kinetic Translational Energy, Transformic Translational Energy, and Potential Energy. See Figure 9 and Appendix.

TRANSLATIONAL ENERGY AND ROTATIONAL ENERGY PARTS OF THE NEW ENERGY EQUATION

Remember in our earlier discussion about motion, that "dynamic" motion includes both translational and rotational velocity? In our examination of the energy associated with motion we need to account for both forms of motion. In every beginning physics class we learn that the energy associated with velocity is called kinetic energy. Also, we learn about another form of energy called "potential" energy.

A popular experiment used in almost every high school or college freshman physics lab is the demonstration of the relationship between kinetic energy and potential energy. This experiment is described in Figures 7A and 7B and uses two separate procedures.

In the first procedure, Figure 7A, we drop a ball from a height of 1 meter and determine the velocity of the ball when it hits the ground. The falling ball has only translational velocity and no rotational energy. The translational velocity of the ball can be considered the velocity of a single "point mass" located at the center of the mass (which is located at the exact center of a ball).

In the second procedure, Figure 7B, we roll the same ball down a ramp with the same height of 1 meter, causing some of the energy to produce rotational, or angular, velocity of the ball. The velocity and energy of the ball at the bottom of the ramp is partly the translational velocity of the point mass, and partly the rotational, or angular, velocity of the ball as it rolls.

Figure 7 - Kinetic Energy from Potential Energy Experiments A and B

Experiment 7A – Dropping a ball to determine kinetic energy of translational (linear) motion

Given: A 10 kg hollow steel ball is at rest at a height of 1 meter above the ground, see Figure 7A.

Question: If the ball is dropped, what is the velocity of the ball when it hits the ground?

Solution:

At 1 meter above the ground and at rest, the ball has full potential energy and zero kinetic energy,

$$E_{potential} = m\, g\, h \qquad E_{kinetic} = \frac{1}{2} m\, v^2 = 0 \qquad \text{where m = mass, g= gravity (9.8m/s²),}$$
$$h = \text{height, } v = \text{velocity}$$

When the ball is dropped and reaches the ground, all of the potential energy has been converted to kinetic energy, therefore,

$$E_{potential} = m\, g\, h = E_{kinetic}$$

$$E_{kinetic} = \frac{1}{2} m\, v^2 = m\, g\, h \quad \text{- for answer to question, solve this equation for v, velocity}$$

$$\frac{1}{2} \cancel{m}\, v^2 = \cancel{m} g h$$

$$v^2 = 2gh$$

$$v = \sqrt{2gh} = \sqrt{2\,(9.8)(1)} = \underline{4.427 \text{ meters/sec}}$$

Since the value of mass is on each side, divide each side by m, it cancels out. This means that the velocity of any object that is dropped from a given height will be the same, no matter how heavy the object is. For example, neglecting wind resistance, a feather will drop at the same velocity as the 10 kilogram ball.

When ball is held at height = h, it has a potential energy of mgh, and zero kinetic energy.

When ball is dropped it accelerates at rate of gravity (9.8m/s²). Acceleration is not dependent on mass or size of object (neglecting wind resistance).

After 1 meter of free fall, $v = 4.427 \; meters/sec$

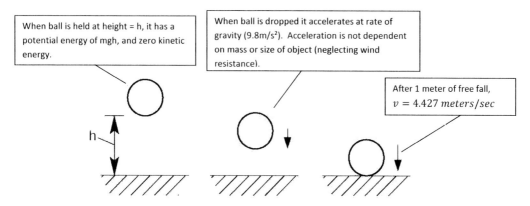

h

Figure 7A – Example of a mass falling with only translational (linear) velocity. Empirical testing, and every day class room experiments, proves the validity of these mathematical equations of energy and the laws of conservation of energy.

Experiment 7B– Rolling same ball down ramp to determine translational and rotational energy

Given: Same 10 kilogram hollow steel ball as in Experiment 7A, is rolled down a ramp that is 1 meter high, and at a 45-degree angle, see Figure 7B.

Question: The ball starts at zero velocity at the top of the ramp, what is the velocity of the ball when it reaches the bottom?

Solution:

The potential energy at the top of the ramp is the same as in Experiment 7A, but in this case, when the ball reaches the bottom of the ramp it will possess a combination of translational (or linear) kinetic energy ($\frac{1}{2}mv^2$) and rotational kinetic energy ($\frac{1}{2}Iw^2$). Therefore,

$$E_{potential} = mgh \qquad E_{kinetic} = \frac{1}{2}mv_t^2 + \frac{1}{2}Iw^2 \qquad \text{where } I = \text{moment of inertia}$$
$$\text{and, } w = \text{angular velocity (like RPM)}$$

When the ball reaches the ground, all of the potential energy has been converted to kinetic energy. The kinetic energy is comprised of both translational and rotational velocity. To avoid complications in calculating I and w for the hollow sphere, we will give the information that the kinetic energy of rotation is 5% of the total kinetic energy, therefore,

$E_{potential} = mgh = (10)(9.8)(1) = 98$ joules

5% of 98 joules for $E_{rotational}$ is 4.9 joules , therefore, $98 - 4.9 = 93.1$ joules is for $E_{translation}$

$E_{translation} = \frac{1}{2}m\,v_t^2 = 93.1$ joules , solve for v, translational velocity,

$v_t^2 = 93.1\,(2)/10$

$v_t = \sqrt{18.62}$ = <u>4.315 meters/sec.</u>

> You can see that the translational (linear) velocity is a bit slower than in Experiment A, however, all of the energy is accounted for by addition of the rotational energy, thereby obeying the Law of Conservation of Energy.

> The ball has a potential energy of mgh, that will be converted to kinetic energy at the bottom of the ramp.

> When ball goes over the edge it will take more time to reach the bottom than ExpA because the travel distance is longer, gravity component is angled, and some of the energy is being transferred to the rotational velocity, w. Regardless, omitting friction and wind resistance, the mgh energy must be conserved and converted to 100% kinetic energy when the ball reaches the bottom of the ramp.

> After 1 meter of h, $v_t = 4.315\ m/s$
> For simplification, value of w not determined

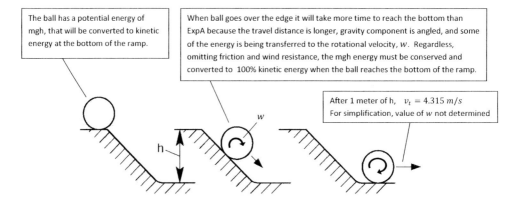

Figure 7B – Example of mass falling with both translational (linear) and rotational velocity. The purpose of this exercise is to show that in our calculation of a universal energy equation, we must account for the rotational energy component, in addition to the translational energy, for all matter in the Spiroverse.

These equations can easily be verified in a lab. Again, it is a common exercise in a beginning physics class to conduct this experiment and verify the calculated results to the measured results. The results of experimentation show that these equations are true and absolute.

We can observe a lot from our two examples in Figures 7A and 7B. It should impress on you that we know a lot about what energy is and what it does on a macroscopic level. The mathematical principles shown here are what enable us to put satellites in orbit, send rockets to the moon, and much much more.

The two examples in Figure 7 were provided to show that, for our energy equation derived in Figure 6 to be complete, we need to account for energy associated with the angular velocity, or rotation, of the mass in the universe. Clearly, the motion in the universe is not solely translational, or linear. By the very nature of the Spirogrid, everything in the universe has a component of rotational velocity. The effects of this rotational velocity, or spin, can have profound effects on the energy associated with matter, and more specifically with the rate of chemical reactions, as described in The FNU Ongoing Wikappendix.

THE TRUE MEANING OF "MOMENT OF INERTIA"

Before we show how we account for the energy associated with rotational, or angular, velocity of all matter in the universe, it is necessary to understand what the moment of inertia, I, really means.

As mentioned before, in today's world of engineering mechanics, including the study of statics and dynamics, the term *inertia* has no meaning on its own. The correct term to describe the energy of a mass in motion is *momentum* or *kinetic energy*. However, every object or mass has a *moment of inertia*.

The moment of inertia, I, is an important and well-known quantity. The moment of inertia, with the units of mass times distance squared (kg-meters²), is a value that quantifies how the mass of an object is distributed around a specified axis (whether the object is rotating or not). Usually, to determine the value of I for an object, an engineer refers to a standardized list, or table, of equations that includes all the common shapes, such as spheres, cylinders, discs, hoops, rods, I-beams, etc. These equations have been derived over the centuries since the Industrial Revolution began.

In its most basic form, the equation for the moment of inertia of a point mass at a distance, r, from the axis of rotation is exactly:

$$I = moment\ of\ Inertia\ (point\ mass) = mr^2$$

where m = point mass m (kg) and r = distance from axis of rotation (meters).

For a solid sphere, like a steel ball, the moment of inertia is exactly:

$$I = moment\ of\ Inertia\ (solid\ sphere\ shape) = \frac{2mr^2}{5}$$

where m = mass (kg) of the sphere and r = radius (meters) of the sphere. Easy.

The classic example of the effect of the moment of inertia is an ice skater that swings her arms out wide in a turn to generate a spinning motion, and then pulls her arms in, close to her center, even straight up over her head, to bring all her mass as close to the axis of rotation as possible. Pulling her arms in close lowers the value of the moment of inertia, causing her rotational velocity to increase, in obeisance to the law of conservation of energy.

We will prove in Figure 8, coming up next, that the equation that defines her rotational energy is very similar to what we determined in Figure 6 for translational energy, and is exactly:

$$E = Rotational\ Energy = \frac{1}{2}Iw^2$$

where E = energy (joules), I = moment of inertia (kg/m²), and ω = angular velocity (radians per second, similar to RPM—revolutions per minute).

In the example of the skater, her arms are out at the beginning of her spin (I_1), and she is rotating at an angular velocity of w_1. At the end of the spin, her arms are in (I_2), and she is rotating at an angular velocity of w_2.

However, due to the conservations of energy, during both conditions she has the same rotational energy (E). Note, for this example, we are omitting the work necessary to bring in the arms.

We illustrate this mathematically as $E_1 = E_2$, and, using the equation for rotational energy (E) described above, we have:

$$E_1 = E_2$$

$$\frac{1}{2} I_1 w_1^2 = \frac{1}{2} I_2 w_2^2, \quad \text{solving for the value of } w_2 \text{, we get}$$

$$w_2 = w_1 \sqrt{I_1 / I_2}$$

Substituting the values for a point mass for each of the values of I, where $I_1 = mr_1^2$ and $I_2 = mr_2^2$, we have:

$$w_2 = w_1 \sqrt{mr_1^2 / mr_2^2} = w_1 \cdot \frac{r_1}{r_2}.$$

Notice that the mass cancels out!

The fact that the mass cancels out tells us that this is an intrinsic characteristic of the rotating system, independent of the material. The change in rotational energy is solely a function of the geometry of the mass, which applies to all mass of all sizes, including the atomic makeup of matter itself. Please note that this equation specifically applies to a point mass, but with maybe a few exceptions, regardless of what geometric shape was used, the mass component of the value of I would cancel out, and only the ratio of the r values would change.

Simply said, the equation shows us that when you lower the value of r_2 (pull arms in), you speed up (increase the value of w_2). Vice versa, also, when you increase the value of r_2 (let arms out), you slow down (decrease the value of w_2). And this is exactly what happens to the rotating ice skater.

Understanding of the effect of a changing moment of inertia, ΔI, is important for our effort to harness the energy of the Spirogrid. But, before we go into that, we need to get our science in order. Figure 8 shows how we derive the rotational energy component for our New Energy Equation, and Figure 9 is a recap that combines both the translational and rotational energy components into the complete New Energy Equation.

Figure 8 - The New Energy Equation for Mass with Angular Moment of Inertia

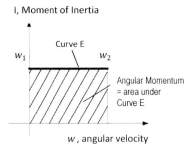

I, Moment of Inertia

Curve E

w_1 w_2

Angular Momentum = area under Curve E

w, angular velocity

With reference to Curve E (actually a line in this case), using the modern expression for Integration, we can write:

$$\text{Angular Momentum} = \int_{w1}^{w2} I\, dw = \text{Area Under Curve E}$$

This equation says: "the summation of an infinite number of tiny rectangles between the value of w_1 and w_2 equals the exact area under the curve, which equals the Angular Momentum of the object".

The result of this integration is:

$$\text{Angular Momentum} = M_I = Iw + M_{Io}$$

Where, I = moment of Inertia for mass (function of mass and geometry, units of kg-meter²), w = angular velocity (units of radians per second), and M_{Io} = angular momentum before the objects angular momentum was changed, which is the constant of integration that must automatically be added, see Note E.

Note E: As in prior examples, M_{Io} accounts for the possibility that the object may have already been rotating before the change in w_1 occurred.

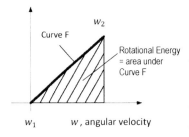

M_I, angular momentum

w_2

Curve F

Rotational Energy = area under Curve F

w_1 w, angular velocity

With reference to Curve F, using the modern expression for Integration, we can write:

$$\text{Energy} = \int_{w1}^{w2} M_I\, dw = \text{Area Under Curve F}$$

From above, we have,

$$M_I = Iw + M_{Io}, \text{ substituting into equation,}$$

$$\text{Energy} = \int_{w1}^{w2} (Iw + M_{Io})\, dw$$

This equation says: "the summation of an infinite number of tiny rectangles between the value of w_1 and w_2 of the equation $(Iw + M_{Io})$ equals the exact area under Curve F, which equals the exact Rotational Energy of the mass".

The result of this integration is:

$$\text{Total Rotational Energy} = E_I = \frac{1}{2}Iw^2 + M_{Io}w + e_{Io}$$

Where, I = moment of Inertia for mass (function of mass and geometry, units of kg-meter²), and w = present angular velocity mass, and M_{Io} = angular momentum of mass before added angular velocity, e_{Io} = added constant of integration, rotational energy of mass at origin. See Note F.

Note F: e_0 is the absolute origin of rotational energy, before the effects of added angular velocity. See Figures 9 and 10, and Appendix.

Figure 9 – Summation of All Energy in the Universe – with Comments

The two equations, including, Total Tranlational Energy, E_t, from Figure 6, and Total Rotational Energy, E_I, from Figure 8, can be added together, similar to the example provided in Figure 7, to obtain and equation that could account for the total energy that exists in the Spiroverse (Universe), as follows:

Total Energy in the Spiroverse = $E_{Total} = E_t + E_I = [\frac{1}{2}mv^2 + M_{to}v + e_{to}] + [\frac{1}{2}Iw^2 + M_{Io}w + e_{Io}]$

Arranging these into the three (3) fundamental categories of energy, we have:

Total Energy Equation = $E_{Total} = \underbrace{[\frac{1}{2}mv^2 + \frac{1}{2}Iw^2]}_{} + \underbrace{[M_{to}v + M_{Io}w]}_{} + \underbrace{[e_{to} + e_{Io}]}_{}$

Total Kinetic Energy	Total Transformic Energy	Total Potential Energy

Where, m = mass , v = translational velocity, I = moment of inertia of mass, w = angular velocity,
   and,

M_{to}= Translational, or linear, momentum of mass due to spirographic motion
M_{Io}= Angular, or rotational, momentum of mass due to spirographic motion
   and,

e_{to} = Potential energy (translational), source unknown (energy, work, power, mgh, pressure x volume?)
e_{io} = Potential energy (rotational), source unknown (energy, work, power, mgh, pressure x volume?)

Where can "E = mc²" be now?

M_{to} and M_{Io} are both momentums of the mass due to the motion, or velocity, of the spirogridic motion. Remember, these values were added by the rules of integration, and correctly account for the momentum of the mass through space before macro accelerations were made to the mass, see Figures 5, 6 and 7.

Taking the quantity M_{to}, the translational motion of the mass, we make the following equation:

M_{to} = **translational momentum of mass** = (mass) x (velocity of the Spirogrid) = $m\,v_{grid}$

If we substitute this value into the translational velocity portion of the transformic energy part of the "Total Energy Equation", $M_{to}v$, we get the following:

$$E_{transformic\ translational\ only} = m\,v_{grid}\,v_{translational\ velocity}$$

Therefore, if the velocity of the motion of the spirogridic pattern is equal to the speed of light (v_{grid} = c), and the mass was translating, moving through space, at the speed of light ($v_{translational\ veleocity}$ = c), then:

$$E_{transformic\ translational\ velicity\ only} = mc^2$$

This may sound cool, but there are a number of problems with the popular opinions about this equation. The energy of the mass is not intrinsic, as most are led to believe, but is associated with the kinetic energy of the motion of the mass. If such a translational velocity were possible, the E = mc² expression would only be a portion of the total energy of the mass. What proof do we have that v_{grid} = c, maybe it's higher? Many questions arise, including questions about possible limits to the value of translational velocity, and the true relationship of translational verses corporeal motion.

The discovery of the spirographic motion of the universe is what allows us to derive and make sense of the New Total Energy Equation, shown in Figure 9. The understanding of the Spirogrid guides us in our treatment of the otherwise unknown and ambiguous variables and constants. You can now understand why this knowledge of the Spirogrid is referred to as "the Fundamental Nature of the Universe."

For example, if you read Handout A, which provides background on the derivation of Einstein's energy equation, you will see how there is ambiguity regarding the placement of variables and the use of the speed of light in a vacuum.

The constant associated with the complete New Total Energy Equation, v_{grid}, is associated with the kinetic energy caused by the spirographic motion of the universe. This is much easier to associate with energy, compared to "c," the speed of light in a vacuum.

There has been too much blind acceptance of using the speed of light in a vacuum for a universal energy equation. What is the connection again? I never really got it. Can a true vacuum really exist? Current understanding of particle physics including neutrinos and the simple cloud chamber experiment show us that there are particles constantly flying straight through us. How can a true vacuum exist if particles are always flying through everything? What about hydrogen permeability? How could there ever be a true vacuum if every material is permeable to hydrogen? Would not the hydrogen atoms diffuse into a vacuum just like water molecules through a semi-permeable membrane of a reverse-osmosis filter? The fact of the matter is, there is no basis to the claim of the speed of light in a vacuum, simply because is has never been confirmed by experimentation.

In our derivation of the complete New Energy Equation, we make the mass associated with the energy of matter a constant. This is in agreement with the observation of the law of conservation of matter. However, we add a new variable to the energy equation, namely I, the moment of inertia, which we know changes during chemical reactions. We know this because chemical reactions result in a change in density of the reactants versus the products of the reaction.

The association of the moment of inertia to the newly defined transformic energy portion of the New Total Energy Equation is probably the most significant breakthrough associated with the discovery of the Spirogrid. This relationship is described in Figure 10 and involves thermal energy (temperature), changes in density (ρ), and the energy associated with chemical reactions.

This discovery of transformic energy, ladies and gentlemen, is where we are going to make our money.

Figure 10 – Transformic Energy with Changing "I" – with Comments

From Figure 9 we show the New Total Energy Equation as:

Total Energy Equation = E_{Total} = $\left[\frac{1}{2}mv^2 + \frac{1}{2}Iw^2\right] + [M_{to}v + M_{Io}w] + [e_{to} + e_{Io}]$

| Total Kinetic Energy | Total Transformic Energy | Total Potential Energy |

Where, m = mass , v = translational velocity, I = moment of inertia, w = angular velocity ("rpm"),
 and,

M_{to}= Translational momentum of mass due to spirographic motion = mv_{grid} (see Fig. 9)
M_{Io}= Rotational momentum of mass due to spirographic motion = $I_{Io}w_{Io}$ (see below)
 and,

e_{to} = Potential energy (translational), source unknown (energy, work, power, mgh, pressure x volume?)
e_{Io} = Potential energy (rotational), source unknown (energy, work, power, mgh, pressure x volume?)

Transformic Energy, $M_{Io}w$ - Effect of changing I_{Io} - The true "intrinsic" Energy of Mass

$M_{to}$and $M_{Io}$are both momentums of the mass due to the motion, or velocity, of the spirogridic motion. In Figure 9 we showed how E=mc² could be derived from the translational momentum, M_{to}, of transformic energy (though meaningless). A more significant expression for intrinsic energy of mass can be derived from the intrinsic angular momentum component, M_{Io}, of transformic energy.

Taking the quantity M_{Io}, we make the following equation for the value of angular momentum:

$$M_{Io} = \text{angular (rotational) momentum of mass} = I_{Io}w_{Io}$$

Where I_{Io} = gridic momentum of inertia of the mass, and, w_{Io} = gridic angular velocity of mass

If we substitute this value into the rotational velocity portion of the transformic energy part of the "Total Energy Equation", $M_{Io}w$, we get the following:

$$E_{transformic\ rotational\ only} = I_{Io}w_{Io}\ w$$

If we apply this equation to the atomic structure, we have tremendous insights into the intrinsic chemical and thermal energy of mass. We could go a step further and suggest that this equation *is* the intrinsic energy of matter.

It is important to note that each of the factors in this equation are variable. For example, thermal energy, the energy that makes an object warm or cold, is related to the value of the rotational velocity within the atomic structure of the mass, w_{Io}, that changes with temperature.

Most interesting is the effect of a change in value of I_{Io}. For example, if we use the value of moment of inertia for a point mass at a distance r from axis or rotation, = mr^2 , and insert it our equation, we get the following:

$$E_{transformic\ rotational\ only} = I_{Io}w_{Io}\ w = [mr^2]w_{Io}\ w$$

We can see how this equation could be applied considering the fact that chemical reactions, such as the combustion of hydrocarbons, cause a change in "structure" of the matter, including I_{Io} and density (Δρ), between the reactants and the products of the chemical reaction. This equation says that a change in I_{Io}, results in a change in energy. Furthermore, a chemical reaction causing a decrease in I_{Io} (shorter "r"and higher w_{Io}) is exothermic, and one causing an increase in I_{Io} (increase "r"and lower w_{Io}) is endothermic. Therefore, in consideration of this discovery, it is apparent that nuclear, chemical, and thermal reactions are defined by the equation for transformic energy.

As mentioned before, attached to the end of the handout is an appendix. It is called "The Fundamental Nature of the Universe (FNU) Ongoing Wikappendix." Explanations for some of the questions raised here are provided, including a description of how the media of space is not a true vacuum but is in fact diatomic hydrogen at extremely low pressures (a vacuum is a relative concept), and how light is only a wave and there is no photon or wave-particle duality.

Additional work will be conducted to determine the effects of dynamic versus corporeal motion on the FNU energy equations, and additional work will be conducted to determine the true meaning of the transformic and potential energy components of the complete New Energy Equations. Further descriptions and discussions are part of The FNU Ongoing Wikappendix.

In The FNU Ongoing Wikappendix, a new model for the atomic structure is proposed, called the Vocuometric Atomic Cell (VAC). This model is based on the new understandings of energy, the atomic moment of inertia, the motion of the Spirographic Grid pattern in the universe, and the New Total Energy Equation.

****
End of Manta's Lecture
****

This material concluded Manta's presentation for the day. Despite the intermittent breaks for food and drinks, the group sat in the conference room all day and listened to the concepts their fellow group member walked them through. Manta glanced at the clock and saw he finished just in time for a group discussion. He opened up the floor, "Any questions?"

After a period of silence, a mood of nostalgia began permeating the group. They felt like they were back in high school, curious and learning fresh material. However, as soon as the air cleared, they were back to their CEO instincts.

"Yes, I have a question," said Carlos from near the back of the table. "You have mentioned before how we are going to make money on this discovery. What specifically can you tell us about how we are going to do it?"

"In time, my friend," said Manta. "We still have one more meeting, which I would like to conduct tomorrow, if that is in favor with everyone. Let me give you this clue, though. There is a rule in nature that when you compress a gas, it warms, and when you decompress a gas, it cools. There are a number of gas laws that describe this phenomenon, including the Ideal Gas Law. If you look at the change in temperature being created, you will see that it relates to the change in the moment of inertia of the atomic structure of the molecules of gas," explained Manta with the aid of the whiteboard.

Manta continued, "As described in Figure 10, when you change the moment of inertia of a gas, the temperature of a gas changes. The gas temperature will increase with increasing pressure (I_{10}, shorter "r" and higher w_{10}) and the gas temperature will decrease with a lower gas pressure. So here is your clue. The effect of changing the moment of inertia also applies to electrochemical reactions. Instead of changing temperature by compressing a gas, we are going to change the direction something moves. So, basically, we are going to be in the moving business—the business of moving electrons.

"That's all I'll say until the next meeting," finished Manta.

"Ask a technical question and you'll get a technical answer, and, not surprisingly, in the form of a riddle," said Carlos, getting the whole group to laugh.

"Especially when the person is Miles Manta," added Bill over the laughter in the room.

"OK, are we all in favor of meeting again for Miles's third disclosure tomorrow, like he suggested?" Bill asked the group.

The vote was a resonating "aye" and Bill adjourned the meeting. The group chatted about dinner and family plans, also about how they were keeping abreast of their business activities via satellite and email communication. All of the members except Bill had left the conference room.

Manta let Bill know he was taking the flash drive recording of the meeting.

"You want to join some of us for dinner tonight downtown?" Bill asked Manta.

"Thank you sincerely for the invite, but I better get home to my family. I want to help my oldest son with a school project tonight," Manta replied.

"Maybe next time," Bill replied, smiling. "OK, well, I'm off. Tell your family 'hi' for me."

Manta was grateful for a friend like Bill.

He closed down the room and left the convention center feeling refreshed after how well the progress went today in the meeting. On his way home, he rolled down the windows to breathe in the fresh air. He definitely enjoyed feeling less of the burden these past few days. Tomorrow would be a very important day for him—the third meeting—but first things first, his mind began focusing on how he could help his son with his school project tonight.

CHAPTER 8

SCIENCE, FASHION, AND THE ETHER

Reports about the atomic disaster were still headline news. Every major channel was swarmed by more details, facts, figures, and theories. There seemed to be less blaming going on, but, still, no one had claimed responsibility, so there was a wide window for suspicion. This was exactly how Manta had predicted it.

Plenty of conspiracy theories buzzed around. Why did this tragedy have to strike right when things were winding down in Afghanistan? Defense spending cuts were being reversed. And of course, in response to this new threat, new agencies were already added to the US government. People were skeptical about the bills being rammed through Congress and were questioning the loss of more rights and liberties.

TREK MEETING NUMBER THREE

All twelve members met in the same conference room on Wednesday morning. Some of them discussed different news reports they had heard and which theories they thought had more credibility about who could have possibly set the bomb off. Others enjoyed the coffee and muffins in the back of the room and took to looking out the window to the view of the bay while discussing some lighter issues with each other.

Manta, on the other hand, was quietly setting up his lecture notes and plugging in the flash drive in the computer at the front of the room. He was especially eager to disclose what he had with his group members today.

Manta looked at the clock and asked for everyone to take a seat so he could get started.

Once the shuffling quieted down, Bill Oliver welcomed everyone to the meeting.

"Thank you all for coming. Miles promised us information about an invention. He said it was related to the moving business. I recall my first job as an undergraduate student working for a moving company. My back has never been the same. Maybe the Manta version of the moving business will include an anti-gravity levitation machine. Who knows?" joked Bill. The others laughed.

"We did promise Miles our full attention for the three meetings. Since this is the final meeting, at the end of our session, we will each have an opportunity to express our points of view and announce our intentions of whether we wish to continue our association. We will follow that up again with a final vote. With that, I turn the meeting over to Miles. Let's see what you've got," Bill said.

Manta stood tall, shoulders back, with his hands in his pockets at the front of the room. He had a kind but serious look on his face, and he was dressed sharply in a gray business suit.

"Thank you all for enduring my boring science lessons," Manta joked. "I am confident that all of you will see the value and necessity of understanding the basic working principles of what we are going to be selling," he said as he situated himself behind the lectern and took his hands out of his pockets.

"Our group has been together a long time. As you know, the circumstances that have brought this group together have not been accidental. That may be bothersome to some of you. You may feel that you have been manipulated. I can only emphasize that there is an overriding purpose.

"You are all highly skilled and experienced entrepreneurs who have become outstanding CEOs of your successful corporations. You understand what it takes to get things done, and, most of all, for the sake of what lies ahead of us, you understand what motivates people to do the things they do.

"One of the crowning achievements of the world's technological advancement has been the dissemination of information. The passing of information occurs worldwide and has become almost instantaneous. The good news is that accurate information is available; you just have to find it. The bad aspect of this massive dissemination is the large amount of inaccurate information that must be filtered out. So, in a sense, the amount of work necessary to develop an informed opinion has not changed. Also, the means and methods of persuasion and salesmanship have not changed that much.

"Throughout our recent discussions, I have touched on the events of human history where the masses have either gone in the direction of truth or blindness. If you examine each of these turning points, each case will show that there are elements of greed, ego, power, and, with no better way to put it, fashion.

"To resolve the dangerous issues that this planet will be facing in the future, we will rely heavily on scientific understanding and less on fashion. Frankly, the significant issues we are facing are related to saving ourselves from self-annihilation, either by pollution, consumption, or atomic holocaust.

"There is nothing wrong with religion, politics, or fashion. But when any of these three become the steering force in making decisions, we end up with highly polarized outcomes, usually based on the lowest common denominator or the interests of a select few people or groups.

"Science and fashion go together like religion and politics. Science is about form following function. Fashion is about form following what your neighbor thinks. And we all know the detriment of mixing religion and politics.

"For example, let's say that you have one hundred cups turned upside down on a table. Under some of the cups are little gumballs. Scientifically, the solution to determining the number of gumballs is to systematically lift up each of the cups to determine and count them. It is that simple. But what if someone said, 'ignore these twenty cups over here' because of some reason of religion, politics, or fashion. This situation is completely incompatible with true science. Yet, this is how the world functions, including, unfortunately, the scientific community in some cases."

EXAMPLES OF PETTY FASHION

Manta paused for a drink of water and then continued. "I would like to cite two recent examples that I am certain we all could relate to.

"All of us know about the rock band The Beatles from the 1960s. To say they were game-changers in the world of music would be the understatement of the century.

"But what happened after The Beatles were successful in the US for seven continuous years straight? People started getting sick of them.

"I remember when the last recorded Beatles album was released. The song 'Come Together' was playing on the car radio, and my oldest brother reached down and changed the channel, saying with an expletive, 'I am sick of these guys.' Their album sales were down, and their movie *Let it Be* was a flop.

"If you look at their music, it certainly was not diminishing in quality. On the contrary, their work in exploring and developing new arrangements was showing ongoing improvements, matter of taste of course. Yet when the band broke up in 1969, few seemed to care. At least, that is how it was taken in my neighborhood.

"Another band that comes to mind is from the 1970s. Remember the Bee Gees? They rode the disco craze to the top. They were extremely talented, their music was great—again, matter of taste—and they were poised to do so much more. Yet what happened? All of a sudden they were not fashionable anymore.

"People started destroying Bee Gees albums in the street. If you said that you were a Bee Gees fan, you could be ostracized or judged for doing so! Why? What did they do wrong?

"This human behavior occurred because it became fashionable to hate disco music, symbolized by the Bee Gees. Similar to how it became fashionable to hate The Beatles.

"Of course, over time people realize that they threw the baby out with the bath water, and the true genius of the Beatles and the Bee Gees is recognized—again, a matter of taste.

"It does make life interesting, but we know this is all very petty behavior. Scientists are not immune to petty behavior either."

THE FASHION OF ISAAC NEWTON

As Manta transitioned to the next subject, he passed around an outline to the members so they could more easily follow along

"We have to remind ourselves that the word *science* did not exist during Isaac Newton's time," Manta continued. "He was referred to as a natural philosopher.

"Also, there was no chemistry during his lifetime. People who experimented with the reactions between different materials were called alchemists. The basis for alchemy was the effort to transform ordinary material into gold, and although they were never successful at this, their work helped the science of chemistry advance to what it is today.

"The similarity between Newton's time and today's scientific community is that the government—the kings and queens then, and grants and other programs now—provides support and funding for new scientific discoveries.

"Try to take yourselves back in time. What would be the criteria for whether you received government support or not? Clearly, new weapon development would be something they would fund. The effort to turn ordinary materials into gold would be very attractive. Or the possibility of better gunpowder would receive their support. Anything else come to mind?

"Remember, Newton's discoveries describing how the planets rotate around the sun may not have had great appeal to kings and queens; what was the benefit, especially if it contradicted what the church had to say? The use of his mathematics to project a cannon ball would generate interest, but how else could Newton sell his ideas?

"Just like today, in Newton's time there was great entertainment value in natural philosophy. It was fashionable at the time for higher society to attend lectures at the educational institutions. Tickets for popular lecturers were highly sought after, similar to today's music celebrities.

"Therefore, if you had a discovery that had entertainment value, you would be in a better position to impress the king or queen. For example, the person who invented the Leyden jar could store static electricity in a lead-lined glass jar. Coercing some unsuspecting soul into touching both terminals would release a jolt of very high voltage, causing them to scream in agony. Now *this* would get the king's attention and would be worth seeing again. Maybe government funding hasn't changed that much after all?"

Some members laughed under their breath realizing the truth of that statement.

Manta continued. "Newton was knighted by Queen Anne in 1705, at the age of sixty-three. With the success of colonial expansion and the beginnings of the Industrial Revolution, Britain was riding high on great wealth. Newton was a national hero. Even after his death, Newton was growing in legendary status.

"His fame and the loyal following of his theories presented in the *Principia* publications were similar to what we see today in adherence to Einstein's theory of relativity. Isaac Newton was a megastar."

NEWTON'S OTHER WORK—LUMINIFEROUS ETHER AND NEWTON'S FALL FROM FASHION

Manta asked if they needed a short break. Everyone seemed to want him to continue, so he carried on. He pointed out that if anyone needed a break just head to the back of the room, and the rest of the group would take that opportunity to stretch.

Manta continued, "Besides Newton's definition of forces that affect bodies of mass and his invention of calculus, another branch of his work involved the behavior of light and optics. He built the first practical reflecting telescope. He discovered that light was made up of different colors by observing the spectrum that occurred when light traveled through a glass prism.

"He also studied the speed of sound and worked on the principals of how sound traveled as wave energy through a medium. He discovered that sound did not travel through a space devoid of air.

"Logically, he thought, light must travel through a medium just like sound did. Makes a lot of sense, doesn't it? Modern science has conclusively proven that light behaves as a wave, with precisely known frequencies, colors, interference patterns, energy levels, and even what are known as Doppler shifts in frequency when the source is travelling at a different speed from the observer. But what is the medium of light? It was observed that light traveled through a "vacuum", that is, light traveled through what they thought to be a volume void of air or media.

"What Newton proposed was the luminiferous ether as the medium for the propagation of light. This theory was held as a cornerstone of the understanding of the universe throughout the 1800s and just past the turn of the century.

Adherents to one form of an ether theory or another included most, if not all, of the leading scientists of the period.

"These scientists knew that Earth flew through the ether of space at 108,000 km/hr as it followed its orbit around the sun. And if light behaved the same as a sound wave, they thought they should be able to measure a difference in the speed of light on the surface of Earth as it rotated *into* the ether wind, versus rotating *with* the ether wind. The rotation of Earth causes a velocity of 1,670 km/hr at the equator. Therefore, the speed of light should slow down or speed up, so they thought.

"Increasingly more complex experiments were carried out in the latter 1800s that tried to measure the change in the speed of light due to the rotation of Earth. But they failed to measure any difference. The results of one experiment were published in the *American Journal of Science* in 1887, widely known as the Michelson-Morley experiment.

"As you could imagine, discussions about the behavior of light and the ether would have been a popular subject among the science community in the late 1800s and early 1900s. All the big names in science would have been aware of the traditional Newtonian descriptions, as well as other ether models vying for attention within the circle of leading figures in the scientific community.

"For example, some would have promoted the Le Sage theory of gravitation that included a medium of light consisting of unseen tiny particles traveling in all directions throughout the universe. Other examples were the mechanistic models of the ether that relied on imaginative gears and cogs. Some of contributors included James Clerk Maxwell, George Fitzgerald, Oliver Heaviside, and Oliver Lodge.

"Joseph Larmor and Hendrik Lorentz had particle-based theories of the ether, including Lorentz's 'theory of the electrons,' which included models for the propagation of light in space.

Lorentz provided an explanation of why the Michelson-Morley experiment could not detect a difference in the speed of light relative to the rotation of Earth. This explanation was named the Lorentz-Fitzgerald theory and described how the absolute motion of an ether could be not be detectable.

"However, Albert Einstein's theories and the Michelson-Morley results were getting more attention than Lorentz and his theories could generate. Why?

"An examination of the history around the debate reveals a huge opportunity for Albert Einstein to obtain megastar status. Isn't that a powerful motive for most people?

"Einstein was successful in promoting the Michelson-Morley experiment from an average experiment into something that put his theories at the center of attention. Come on folks, this is as much about marketing as it is about science.

"Regarding light, Einstein's theory had three basic components. First, the speed of light was constant in a vacuum. Second, the speed of light was the same to all observers, no matter what speed they were traveling. The third part of his theory was truly unique because he claimed that light did not travel as a wave through space. Instead, Einstein believed light traveled through the vacuum of space by means of a particle called a photon. This wave-particle dual nature, or duality, of light omitted the need for a medium in order for light to travel through space. With this theory there was no need for an ether.

"This is especially convenient because the explanation of a real, tangible, and corporeal ether had become a severe roadblock, similar to the problem of figuring out the area under the curve and instantaneous velocity. Scientists don't like not having an answer. It is like cable television programming today—you have space, so you have to fill it with something, otherwise you just have a blank screen, and you can't market that, can you? Well, maybe you can, having seen some of the latest reality shows".

The group did not respond. "That was a joke. Do we need to take a break?" Manta said, as he managed to get a few chuckles from the group and heads nodding in agreement.

-Manta continued, "With the demise of many attempted ether theories that had failed to satisfy the thorough and scrutinizing examination of the old guards of the Royal Institute, the opportunity to go around the issue must have been very tempting. So, rather than the scientists getting through the problem, they just went around it by changing the parameters. Also, the timing was right for the scientists on the continent to make an advance. George Fitzgerald died suddenly in 1901 and Lord Kelvin died in 1907, which left a vacuum in leadership within the British rein of leaders in the field of physics.

"With the right influence peddling, backing, and nationalistic pride, this was the perfect storm. The continentals had an open opportunity to put their man on top.

"I mentioned nationalistic pride because Einstein was German. Germany had prevailed in the Franco-Prussian War and had a lot of nationalistic—and scientific— momentum. Germany has lagged behind the British during the beginning stages of the industrial revolution. In the eyes of the Germans, and probably the entire European continent scientific community, including the French, Welsh, and Dutch, the British were successful for too long riding on the coattails of Isaac Newton's and Michael Faraday's scientific discoveries.

"As the sun never set on the British flag, the British colonial and industrial expansions were unmatched. Nationalism was the favorite pastime prior to World War I and, no doubt, the scientific community would be caught up in the fervor. Of course there would be rivals among the continentals, but we know what strange bedfellows sports fans make. For example, sports fans will root for their biggest rival if it helps their team in the standings.

"The theory of the ether model was struggling for a breakthrough in understanding or some tangible evidence. With every attack or criticism came a responsible rebuttal or a new set of formulas. This condition led to the famous statement by Lorentz, '…I am at the end of my Latin,' where he simply tired out from battling and counterarguing the existence of the ether versus the fashionistic momentum of the Michelson-Morley experiment.

"The public was growing impatient for the next breakthrough or big hit. The entire European scientific community must have been feeling tired of Newton's success. Yes, everyone thought it was time for the British to be brought down a notch.

"Enter Einstein. He had published a well-received description of Brownian motion, which is the apparent random motion of a molecule suspended in a fluid. He had new and fresh ideas the public enjoyed listening to and reading about. He was an ideal fit for the part as the Continentals' replacement for Isaac Newton. His hair was perfect—it added to the interest and appeal in what he had to say. He was a new and memorable fresh face the public began to love. He was a perfect candidate for megastar status.

"Another great debate that was occurring during this same period of time was about our understanding of the natural world. This debate was related to the work of Charles Darwin and his book *The Origin of Species*. It should be obvious to us that the issues of evolution versus creation were intertwined in the minds of the great and learned men of the science of physics. Darwin's book was met with the similar fan fare of a Harry Potter novel, where his 1st edition of 1,250 copies were sold out in a day and the larger printing of his 2nd edition was sold in a short time.

"Remember, Newton was a highly religious man and a staunch advocate of the Bible. It is known that most of his writings were religious dissertations. Advocates of the British and Lorentz's ether theory would have had the double duty of upholding the contemporary relevance of their old-fashioned ideas, including their religious beliefs, amidst a growing referendum for the separation of church and state.

"How much of the pendulum's swing toward the more secular and agnostic beliefs of Einstein's non-ether theory was due to a straw vote on the Dark-Age behaviors of the church? The church controlled public education during a time when the discipline of teaching was transformed into the art of keeping people ignorant. Certainly, it was fertile ground for the ideas of Karl Marx, who identified religion as the 'opiate of the people.'

"In the end, the popularity of the ether took the same road as The Beatles and the Bee Gees. Not only did the ether theory fall out of fashion, it became fashionable to hate it. The connection of the ether with a belief in God was something highly repulsive to the growing ranks of atheists in high levels of science, politics, and society—including men like Sir Arthur Conan Doyle, the popular author of the Sherlock Holmes mysteries, who was accused of being the perpetrator of the Piltdown Man hoax of a forged human fossil that took more than forty years to uncover.

"The theory of the ether died overnight.

"Einstein's theory of special relativity grew into what is called quantum mechanics. There should be great sympathy for any person that embarks upon the understanding of quantum mechanics," Manta said putting his hand on the side of his face and cringing.

"I don't think that anyone would argue too much against the suggestion that quantum mechanics has become a legitimized form of insanity. At its core are beliefs of esoteric concepts with a fixation on a counting number sequence, quanta, that seems to morph on a weekly basis. Quantum theories have become very similar to complicated religious concepts that require huge amounts of faith in the unknown, or unknowable, which are understood only by a select priesthood. It seems that if you can't explain something with real understanding, just delegate it to the field of quantum mechanics where you can impress someone with your articulate ability to say you have no clue what is happening, without consequences.

"We have painted a picture where the advocates of the ether hit a dead end and were frustrated. Yet, hitting dead ends is part of the journey to discovery, just like the area under the curve problem. The frustration encountered by the advocates of the ether theory should not be interpreted as a failure.

"A large problem with the quantum-mechanic advocates is that they irresponsibly plow through dead ends with more layers of 'Latin.' Except, in this case, it is really not Latin, it is 'Martian.' If you don't believe me, try listening to Stephen Hawking talk to his interpreter," Manta joked.

There was a mixed reaction from the group, some laughed at the Martian remark, others cringed as though it was irreverent.

"This issue of the ether and the medium of light are going to be the most contentious issues that we will face in our disclosure of the Spirogrid concept. They will be the most significant obstacles in gaining acceptance. The adherents of Einstein will explain away anything that contradicts their belief with their exclusive quantum language that only they can interpret, if it can even be interpreted at all."

With that, Manta suggested the group break for lunch. He could tell the group was listening attentively and was very intrigued by his presentation, but he knew they could use some fresh air and refreshments.

The group once again converged on the nicely catered meal that had been prepared down the hall. Manta didn't join the group for lunch, but he left the room as well. He took a slow paced walk outside around the convention center to get some fresh air and found a quite spot to enjoy a light lunch.

After thirty minutes, they all met back in the conference room.

AN ETHER MODEL—NECESSARY OR NOT

"You have all been very good listeners. Thank you," Manta began as the last of them were taking their seats and getting comfortable.

"Let's continue. Shall we?" Manta asked the group, rhetorically. He flipped a page in his notes and jumped in again.

"I think you all know how I feel about the state of the scientific community regarding the physical makeup of the universe. I am concerned that the immediate rejection of an ether model will be fierce, and the spokesmen of modern science, namely the TV personalities, will stay dedicated to their quantum models in the same way that a religious person bows down to a statue or idol.

"Time has shown us that the greatest scientists are those who cautiously describe what they think they know, and acknowledge that they really know very little. An examination of the greatest scientists on record would reveal that they were frustrated by their lack of understanding instead of being proud of their accomplishments.

"I want to point this out because, in our effort to get through this material describing the Spirogrid, we will rely on models of motion. For our discussions there are two specific forms of motion—dynamic motion and corporeal motion. Each form of motion is associated with different energy and force characteristics. When we examine specific natural phenomena, we may not know the extent of these force interactions. Therefore, our models may be limited.

"The models based on the dynamic motion of the Spirogrid explain many phenomena, including our example of Drew's kinetic energy, and the example of the lower 'high' tide as the earth rotates about the barycenter. How much more can be explained with only the model of the dynamic motion of the Spirogrid? Remember, this would be attractive to a lot of scientists because there is no requirement for corporeal ether.

"On the other hand, there are several reasons why we want to resurrect the concept of the ether. One is that it was never really disproved, it was only pushed aside as being superfluous. In the search to discover the ether, the Michelson-Morley experiment does not apply for many reasons, as we will show in the FNU Wikappendix. New models of the universe provided by the discovery of the Spirogrid will allow the development of new methods of detection and experimentation that that will prove the existence of the ether. Together with the Spirogrid, the corporeal ether provides a powerful model, or tool, to explain natural phenomena, such as gravity, strong nuclear forces, and electromagnetic induction.

"One of the most important tools in science is the development of models that represent the unseen. For example, we should all be familiar with the Bohr model of the atom, which shows a nucleus surrounded by the layers of electrons, with the nucleus made up of equal amounts of protons and neutrons.

"The Bohr model of the atom has been used extensively to describe the behavior of chemical reactions and the formation of compounds. But, how accurately does it really describe the physical makeup of matter? Are we really supposed to believe that a water molecule looks like Mickey Mouse? Of course not, the model of H_2O only helps us understand the concept of the molecule and we do not really know what the exact physical appearance of a water molecule looks like.

"Another model that can be overstated is the photon nature of light. The wave-particle duality of light, and the concept of the photon, are only models that attempt to explain why light travels through a vacuum.

"As described in The FNU Wikappendix, a true vacuum cannot exist in the macroscopic realm. Also, the long-sought-after medium of space through which light waves travel is not the ether; it is simply diatomic hydrogen at an extremely low pressure. Light is an electromagnetic wave that travels through any transparent medium, including the diatomic hydrogen medium of space, and there is no wave-particle duality.

"Before we delve into our model of the ether, it would be helpful to review an event in the past that has a high level of scientific significance today, but is a hundred and eighty degrees from the truth. But, before I do so, I would like to take a short break. Feel free to get up and stretch your legs, if you wish," Manta said.

BENJAMIN FRANKLIN'S WRONG GUESS

Once they all were seated again, Manta took a deep breath and continued.

"So, as I was saying, I want to review with you a scientific discovery that was incorrect, but which holds a lot of significance.

"In the late 1700s, a prominent American statesman, Mr. Benjamin Franklin, was experimenting with the latest craze of static electricity. At the time, people knew very little about this subject. Some of the tools of the trade that Franklin had at his disposal were pieces of amber. As all of you would know, amber is the hard material that forms from tree sap over thousands of years, and resembles hard plastic, and, by the way, the word for 'amber' in Greek is translated as 'electric,' and all electrical materials during Franklin's time were classified as 'electrics.' He also used a sulfur globe, rabbit fur, wire conductors, and a Leyden jar in his experimentations.

"At this point in time, the only type of electricity that could be used for experiments was static electricity, the same type of electricity that forms on dry days from your shoes rubbing the carpet. Experiments included creating arcs by rubbing the amber or sulfur globe with the rabbit fur, picking up pieces of paper with a static charge, and charging the Leyden jar with a circuit from the sulfur globe. A Leyden jar is like a primitive capacitor that can store a charge of static electricity.

"One of Franklin's experiments was the famous kite experiment, where he was able to charge a Leyden jar with electricity from a strike of lightning. The key that he attached to the end of the string was the center electrode path to the Leyden jar. This experiment proved that lightning was the same substance as what was derived from rubbing the sulfur globe. This was a significant discovery, considering there were theories at the time suggesting that electricity and lightning was a living organism, such as Luigi Galvani's prominent animal electricity theory."

Some of the group members were jotting down notes on the outline he had passed around at the beginning of his lecture. After a brief pause by Manta they drew their attention back to him, and he continued right along with his material.

"The kite experiment made Franklin famous amongst the scientific circles. He continued experimenting with electricity, and, having gained the respect of the scientific community and megastar status, his opinions carried a lot of weight.

"Another significant contribution Franklin made was his identification of the positive and negative terminals of an electrical circuit. He suggested that when matter contained too little of the fluid, it was negatively charged, and, when it had an excess, it was positively charged. When the electrical fluid flowed through the circuit, it traveled from the positive , or abundant, matter to the negative, or deficient, matter. He identified the positive terminal with the vitreous portion of the circuit, and the negative side of the circuit with the resinous terminal. So, the next time you look at a battery and see the plus and minus signs, you can think of Benjamin Franklin.

"During this period of time, there were various branches of science all touting their significance and contributions to the explanations of natural phenomena. One particularly influential group was the scientists developing models of electromagnetism and electrical induction. They used Franklin's positive and negative convention in their experimentation of electric fields. Some of the influential people in this group included: James Clerk Maxwell, George Fitzgerald, Oliver Lodge, Oliver Heaviside, and Lord Kelvin. Of course, Michael Faraday would have also set the early framework for the discipline of magnetism and electrical induction.

"As the science of electrochemistry was further developed in the latter nineteenth century and into the early twentieth century, clear understandings of the specific chemical reactions occurring on the electrodes of electrochemical cells, or batteries, helped scientists discover that the electrons in an electrochemical cell flow in the exact opposite direction than Franklin had suggested. This is the important part: despite this discovery, to this day, it is a fact of scientific convention that electrical "current" flow is opposite to the "electron" flow in a DC circuit.

"In other words, instead of the scientific community joining together and saying, 'OK, we were wrong, let's make the appropriate change,' the electrical induction advocates said, 'Forget that,' and would not change their position, and insisted on maintaining Franklin's convention of current flow.

"Probably because of the timid nature of the electrochemists, with their boring experimental procedures and test tubes, they could not overcome the momentum and influence of the highly fashionable, sparky, and glitzy theoretical work of the Maxwellians.

"Instead of declaring Franklin wrong and making a simple correction to make everything consistent, the scientific community accepted his error as part of their creed. To this day, when you purchase a metering device that measures DC electrical current, including a common everyday multi-meter, it will show the positive flow of current to be in the opposite direction of electron flow. Also, the phenomenon of electrical induction with the right-hand rule is completely opposite from the true direction of electron flow. In another slight to those lefties out there, we should really state the left-hand rule to convey the direction of the induced field from the flow of electrons.

"I am telling you this story to point out that scientific progress is not dependent on every theory being puritanically correct. In many cases the benchmark, or common standard, that the models provide allows us to form the basis for our work of discovery, and effectively communicate our ideas. Although it would be ideal to be exactly correct and consistent in every branch of scientific endeavor, what is at least as important is that scientific models are based on theories to help us 'see' what is invisible.

"The model of the Spirogrid, the new Total Energy Equation, and the recognition of dynamic and corporeal motion combine to provide us with amazingly powerful tools to develop additional models and theories about the makeup of the physical universe.

"Now I would like to take a turn in our conversation to the subject of the ether."

THE ELECTRON COMPONENT OF THE ETHER

Manta looked out the window for a moment, and then he focused back on the members of the group.

Manta continued, "The idea of the electron being part of the elusive ether is not new. The electron media was the cornerstone of Hendrik Lorentz's ether model—the leading ether theory before Einstein's relativity came into vogue. Lorentz received the Nobel Prize in 1902, and the Nobel Lecture on that occasion was titled 'The Theory of Electrons and the Propagation of Light,' where the theory describes 'the physical world as consisting of three separate things, composed of three types of building material: first ordinary tangible or ponderable matter, second electrons, and third ether.'

"The electron is a fascinating discovery. In many senses, its behavior is very simple to understand. In basic direct current (DC) electronic and power applications, the electron behaves perfectly analogous to water flowing in a pipe. This relationship of water flow to electron flow includes electrical voltage being analogous to water pressure in a pipe, electrical current being analogous to water flow in a pipe, and the resistance of electrical current flow being analogous to the resistance of water flow in a pipe.

"We can 'see' the electron with very simple demonstrations of arcs of static electricity, light emissions in a cathode ray tube, and, as Franklin proved, a bolt of lightning. We can also use it to do work, measure it, change its course, and even play with it. We do these things every day.

"What we know about the electron, and how it may possibly be related to an ether substance, is part of The FNU Wikappendix.

"The FNU Wikappendix provides a model of the electron that we know and use today as the incompressible liquid phase of a subatomic substance. According to this model, the long-sought-after ether medium is the compressible gaseous form of the electron, named the 'ethertron,' which is the apparent corporeal component of the Spirogrid.

"Please note that the ethertron is not proposed as the medium through which light travels through space. Again, the FNU Wikappendix describes the medium in space for light to travel as simply diatomic hydrogen at extremely low pressure. In other words, the majority of outer space is not a vacuum. Light does not propagate through a true vacuum, and the notorious black holes of outer space, and the concept of dark matter, could be regions where true vacuums exist. Also, due to the variance in ethertron density, gravitational forces are not constant throughout the universe. "

THE BOHR MODEL OF THE ATOM AND MOLAR VOLUME

"Currently, what is traditionally taught in the schools is the Bohr model of the atomic structure. This model describes the atomic structure as having a core of protons and neutrons, with an outer formation of electrons circling around the core. For the benefit of advancing our understanding of the atomic structure, a new model is necessary.

"Why is a new model necessary? Because the Bohr model does not relate in any way to many observed material phenomena. For example, the Bohr model has no mechanistic connection to phase changes (solid to liquid, liquid to gas), latent heats of fusion and vaporization, temperature, thermal conductivity, magnetism, strong nuclear forces, weak nuclear forces, electromotive potential, molar volume, and others.

"In other words, the Bohr model is terrible." Manta said with an exasperated sigh.

"The Bohr model was based on an experiment of shooting neutrons through a thin foil of gold, and observing that, occasionally, a neutron bounced straight back at the source. Well, this could be explained in a lot of ways. Why should this observation be the primary consideration for the structure of the atom, in deference to the large number of the other phenomena of material matter?

"One of the main problems of the Bohr model is the observation of molar volume. Of all the observed phenomena in the material world, the model of the atomic structure should provide a mechanistic explanation for molar volume."

Manta paused and examined the faces of the group. They were patient listeners, but he was concerned that he was losing them. Manta reminded them that the purpose of this discussion was to familiarize them with an exclusive technology that they could commercialize. It was essential that they understood the fundamentals of how it worked.

Manta pressed on. "When trying to build a model of the atomic structure, or the ether, it is best to look for clues in the physical makeup of nature. Usually these clues occur in the form of unusual or unexplained phenomena in the behavior of matter. One such clue is the existence of molar volume.

"Some of us here have heard of the term 'mole' from a chemistry course. Maybe it gives you chills remembering having to cram for a midterm exam. However, most of you probably are not familiar with how the mole was discovered, or what led the famous scientist Amadeo Avogadro to his work regarding the quantity of a 'mole' in the late eighteenth and early nineteenth centuries.

"Like most discoveries, what led him on was a non-intuitive oddity that defied explanation. He, and others, observed that a certain quantity of gas, whether a pure element or molecules of a compound, and no matter how heavy the element or molecule, always occupied the same gas volume. You have to admit, this is strange. You would think a molecule of argon with twenty times the mass of hydrogen and five times the mass of helium would occupy much more space. But no, they all occupy the same exact volume.

"Again, in as few words as possible, according to the phenomena of molar volume, one atom or molecule of any gas has the exact volume as any other gas at the same temperature and pressure.

"This molar volume of an ideal gas, which is 22.3334 liters per mole, is a cornerstone of modern chemistry. Chemistry experiments in the decades that followed Avogadro's work discovered that the number of atoms, or molecules, in one molar volume, or mole, of gas is always 6.24×10^{23}. Therefore, one mole, or 6.24×10^{23} molecules of pure hydrogen has the same exact volume of 6.24×10^{23} molecules of carbon dioxide, even though one mole carbon dioxide is twenty-two times heavier than one mole of hydrogen.

"The volume of a mole of a solid or liquid is different from that of a gas. Although a mole of solid or liquid always contains 6.24×10^{23} atoms, or molecules if it is a compound, its volume is obviously much less than a gas, and not constant. The volume of one mole of a solid or liquid varies from one substance to another.

"Again, it is disappointing that the Bohr model of the atomic structure has no mechanistic connection to molar volume.

"In light of the understanding of the energies and motions associated with the Spirogrid, a new model is proposed for the atomic structure that will provide mechanistic descriptions of many physical characteristic of elements and compounds, including molar volume."

VACUOMETRIC ATOMIC CELL AND THE RELEVANCE OF CYMATICS

"As I mentioned, The FNU Wikappendix provides a model describing the electron as the incompressible liquid phase of a subatomic substance. It also describes the ether as the gaseous form of the electron—the ethertron.

"In the Wikappendix, you will find a discussion of a new model of the atomic structure called the Vacuometric Atomic Cell, or 'VAC.' At the core of the VAC is a true vacuum that forms the strong nuclear force that holds the atom together. In this model, in addition to the liquidus electron and the gaseous ethertron, there is a solid phase of the electron called 'speckra.' This model suggests that all of the elements are made up of different sized speckra arranged in a cymatic structure formed by the resonant sound waves, or energy waves, of the Spirogrid.

"The use of the phrase 'sound waves' may be a misnomer for the resonant frequencies of the Spirogrid. However, the term is borrowed from the field of cymatics, where sound waves are used to create shapes and gravity-defying phenomena. The formation of complex patterns, shapes, and dynamic structures using sound waves is one of the most intriguing demonstrations to be found in the laboratory and studios of the science and art community. The uncanny resemblance of cymatic patterns, shapes, and motions to those found in the natural world cannot be coincidental. There is clearly a resonant component to the mass and matter of the Spiroverse.

"In a sense, the cymatic structure model of matter at the atomic scale resembles full-motion action video of plasma flowing through blood veins, where the semi-static cell nature of the blood vessels are only slightly moving in comparison to the rapid flow of the blood cells.

"Each element, and compound, has a specific atomic DNA within its vacuous core that defines its cymatic structure. Each element is made up of a specific size of speckra. The unique characteristic of the speckra is that, regardless of its size, it has a uniform density. Regardless of the size or mass of an atom of an element or a molecule, since it is made up of speckra with the same density and is suspended in a rotating media, it will rotate around at the same distance from the center of rotation. Therefore, when a material is in gaseous form, the subatomic particles of spectra revolve at an equal distance away from the center of the atomic core. This explains why all gases have the same molar volume regardless of the mass of the atom or compound.

"Drawings and discussions of the Vacuometric Atomic Cell are provided in The FNU Wikappendix.

"In conclusion, when describing natural phenomena, each problem must be examined on a case-by-case basis. Are the forces caused by kinetic energies stemming from dynamic or corporeal velocities? Are the energies from translational or rotational motion? Is there a component of potential energy? What are the cymatic shapes, structures, and resonant frequencies? There could be a myriad of combinations and permutations of interrelated factors that make the problem too complex to really understand. But, that is the beauty of it. We can spend a lifetime, or many lifetimes, learning about the physical universe and all the natural phenomena that surround us.

"This concludes the theory and science part of my disclosure," said Manta, moving to the side of the podium.

"All the theories and science were necessary to give you a technical foundation before the advent of the products that we will offer to the world.

"Although we have been through some long lectures, I know you are not the rabbit-out-of-the-hat type of business people, and you would want to know the nuts and bolts of the technology.

"Now it's time to reveal what you have been waiting to see," said Manta excitedly.

He walked to the back of the conference room and opened the door to a small walk-in closet. He disappeared into the closet for a moment and then emerged with a stack of bound volumes of documents. He handed them out, passing one copy to each group member. The documents had "Business Plan" written across the cover. He disappeared into the closet again to grab the rest of the documents to pass out to the group. To say these bound documents were hefty would be an understatement. They were three-inches thick and jam packed with information. The group didn't know what to think yet, but Manta, on the other hand, was very excited this moment had finally come—his big revelation.

He disappeared once more into the closet. This time he reemerged pushing a cart with a cover draped over the top.

CHAPTER 9

BREAKTHROUGH TECHNOLOGY AND
THE BUSINESS PLAN

Miles Manta pushed the cart to the front of the conference room. He pulled the cover off and revealed what looked like a shiny, polished stainless steel contraption. There were no apparent moving parts, only a couple of switches and an outlet plug.

"Before we go into the Business Plan I just passed out to you, I wanted to show you the technology, because I know that you have all been waiting for the game-changing invention that I have been alluding to," said Manta as he looked at the group. None of the members were looking back at him, though. They all had their eyes glued to the contraption.

"As I am sure you have all figured out, the key to our technology is the ability to harness energy from the spirographic motion of the Spiroverse.

"Attached to the Business Plan is a patent pending application. This IP covers the key technology to achieve our goal. The basic premise of this invention is that electrical power is extracted from the spirographic motion of the universe by what is called a 'Densiti™' brand of Pressure Density Cell.

"In front of you is one embodiment of this invention. This is a ten-thousand-watt portable system that is about as large as a toaster. This unit should be completely adequate to provide power for an average-sized home, or a small commuter car. They can be added in series, or parallel, to produce practically any power requirement.

"We are set up to have this device made by the millions in a very short order of time. This unit will initially be leased, not sold. All of the financial projections are provided in the Business Plan."

Some of the members were flipping through the Business Plan to see particular sections that interested them. Others were too busy staring at the Pressure Density Cell.

"Yes, there will be some strains by the purveyors of conventional power systems," Manta continued. "But there is a provision in our business plan where they can quickly get their skin into the game.

"After our patent expires there will be healthy competition, similar to the cell phone business today, but our company should be the industry leader, well, for as long as we, or our successors, can make that happen.

"There is really not much more to say about this, other than: Here it is," said Manta.

He hit a switch and a five hundred-watt floodlight shined up to the ceiling. The group was mesmerized.

"Let's take a break and play with this thing for a while," suggested Bill eagerly.

"OK, after the break we will go over some of the key points in the Business Plan," said Manta.

Everyone was out of their seats and gathered around the device.

There was tremendous excitement among the group. Claire became a little emotional and said quietly, "Imagine the underprivileged people in remote parts of the world having simple conveniences such as clean water, lights and refrigeration in their home."

The other group members were thinking of all the possibilities this could potentially have for the world as well.

"Can we make this small enough to power cell phones?" asked Carlos.

"Can we paint some flames on the side of it?" joked Drew, grinning.

"I would like to utilize this in my boat," said Phillip.

They all agreed: this was real, and beautiful.

With the group members watching closely, Miles Manta described how his invention worked.

"The device creates a voltage from a pressure-density gradient. Basically, it is a battery that you pump up with pressure. The pressure, or density, differential creates a voltage difference between plasma interfaces at the anode and cathode regions, just like a battery. But instead of relying on the natural voltage that occurs between two dissimilar metals, the voltage is created by a pressure-density differential caused by the pressure or vacuum," explained Manta.

"Hasn't this been done before?" asked Carl.

"No," replied Manta. "Piezoelectric devices can create a voltage with pressure, but they rely on the electromechanical properties of some materials, such as quartz crystals. They are not related to the design of our pressure density differential invention.

"Any other questions?" Manta asked.

"Yes. Are there any consumables associated with the design?" asked Marcos.

"Yes, there are consumables and ongoing maintenance requirements," Manta explained. "Those issues are addressed in the Business Plan in section seven, I believe.

"And the Business Plan is what we need to discuss next. Let's take a few more minutes to look at this design and then take a seat again."

BUSINESS PLAN—HIDDEN PROFILING

Once all the members had taken a seat, Manta took a seat at the table and began his explanation of the Business Plan.

"I have prepared this formal business plan for you," Manta said, referring to the three-inch-thick spiral bound volume he had provided to each of the group members.

"As I am sure you have done many times in the past with other business plans, you will glance over this material with the full expectation that within a couple of months you will be deviating from the outline and scope. But this plan is necessary, and I think all the schedules, milestones, and expectations are achievable and should not be deviated from too much.

"The basic premise of our plan is based on one question: 'If you had a technology to address the energy needs of the entire world, how would you disseminate the information, build and distribute the technology, and make a profitable and beneficial business for the good of all people and the environment?'

"With that in mind, the objectives can be broken down into four basic categories: Technology, Corporate, Environmental, and Social. These four categories are, of course, highly related and intertwined with each other.

"We already touched on the technology associated with this invention. At this time, I don't think it is necessary to spend too much time on the other aspects of the Business Plan. But there are specific disclosures that I would like to make in what time we have left here today," explained Manta, looking at the clock.

"OK, Miles," interrupted Bill. "Please keep in mind that we are supposed to be voting on our intentions of whether to take on the corporate duties of this new venture or not. I assume that you have predetermined roles and specific duties for all of us, but we first need to vote. And, realize that we need to digest all of this before we vote. Based on your demonstration of the technology, you probably have a blank check with all of our aspirations. But I can only speak for myself."

"Yes, of course," agreed Manta. "The roles for each of you are pretty well spelled out in the Business Plan. We can include that information in our discussions at the end of my presentation, before you vote."

Manta pushed back from table and the materials he had spread out in front of him "You all know the world is on the verge of falling apart. Yes, there is a veneer of cooperation that looks very positive. But there is also an overwhelming sense that everything could fall apart tomorrow. What I am saying is, please keep in mind that we do have to act quickly on our Business Plan for the best intentions of everyone."

Manta stood up and pushed the cart to the side a little so he could have more room to walk behind the lectern, and then he continued.

"The need for a corporate structure is why you are here today. Yes, we will have the technology to change the world, but we could go nowhere without the type of leadership, experience, integrity, and management abilities that you all have demonstrated," Manta explained as he centered himself behind the podium and put his hands in his pockets.

"There is something else that I haven't disclosed to you yet," he continued slowly, with a tongue-in-cheek expression on his face. He hoped they would accept this news well.

Manta saw some group members shift uncomfortably in their chairs and others conjure up a nervous look on their faces. Bill, who was pretty good at remaining the most levelheaded out of the group, was the only one maintaining an unchanged expression.

"Remember when I disclosed to you that I had manipulated the Trek organization, and all of you, when I sought you out to form this group?" Manta asked cautiously. "I hope at this point all of you have gotten over any shock and resentment. And I hope I have regained your trust, because what I have to tell you about next may be just as disconcerting to some of you.

"First, let me tell you about my experience with personality profiling. As CEOs, I know you are familiar with this, but bear with me and allow me to make my point.

"About two years before I joined Trek, I had a lunch meeting with a business associate. I met him at his office. While I was sitting in his office waiting for him to finish up so we could head out, he told me about a personality-profiling service he had retained for his business that helped him interview potential employees. Just for fun, before we left his office, I filled out a simple and straightforward questionnaire, and he faxed it over to the service company.

"After lunch we went back to his office, where the results of my personality profile had been returned and were sitting on his desk. He looked at it and nodded his head in agreement. Then he said, 'Looks like it fits you to a T,' and handed it over to me.

"I read it, and I was completely shocked. It perfectly described me. It was as though someone was peering into my soul. The analysis was so accurate and deep that I felt like yelling out, 'Stop, I surrender.' Of course, being a person of a higher self-confidence, I was proud of the person that the evaluation revealed, both the good and not-so-good qualities.

"In light of my circumstances, trying to figure out who I was and why I was here, I became obsessed with the concept of personality profiling. It was clear to me that this was a method of getting my message out to the world. I quickly became attuned to the science of personality profiling and retained the highest-level experts in the field.

"I realized that I could use these techniques to fulfill a purpose. One of the main reasons I joined Trek was to get a worldwide platform to conduct my personality profile experiments.

"I am sure the light bulb has just lit up in all of your brains. It was I who instituted the mandatory questionnaires within the Trek organization. I instituted it, then they let me commandeer the data, and that information is why all of you are here today and a part of Trek Group T9. I took it a step further, though."

Manta looked from face to face while he was speaking to try to gauge any reactions from the members. He couldn't remember the last time he saw someone move, though. The members were frozen listening to Manta's reveal.

"In a nice way," Manta continued, "I have commandeered all of your companies."

Manta saw a few eyes widen at this point, as he looked around. Walking around the lectern, he pointed as he spoke. "Rajneesh, tell us about that big order that your company has been working on for the conductive ceramic materials for the American chemical company." said Manta.

"Yes," said Rajneesh, "this will be the largest order of these specialty materials we've ever had, maybe anyone has ever had."

"Guess what?" replied Manta. "Those conductive ceramics are the electrode materials for the Pressure Density Cell that is sitting on the table in front of you."

Manta then turned to Carl and asked, "Carl, you know the portable micro-turbine generator project that your company has been developing for that American sports and recreation company?"

"Yes," said Carl, "we have just finished the new factory in Yanzin."

"Well, if you took the cover off that unit sitting on the table over there, it would look very familiar to you, because your company assembled the key components" said Manta.

"OK, Phillip, you're next," said Manta. "Tell us about the German company that hired you to develop the worldwide legal treaty for the electrical power sharing for home-generated solar power."

"That has been practically our whole company-wide effort for the last four years, and it is not limited to solar, but any renewable, home-generated power," Phillip described.

"Yes, that was purely intentional. Your work forms the legal basis of how we can plug the Pressure Density Cells into homes with a simple walk-in permit," Manta replied.

"Oh my goodness," came a voice from the back of the table. It was Malik. "I suppose that huge project for the multi-phase inverters that we have developed over the last four years is how you connect the Pressure Density Cell to the grid?" asked Malik. "And our client had us believing they were for remote water pumping applications."

"I think you are all catching on," replied Manta. "Each one of you have already had a part in bringing this product to the market, and you did not even know it.

"Building and operating a global corporation is not easy, as all of you know. All of you here have followed and improved upon the business models developed by the largest corporations in the world. The Business Plan has been specifically tailored to your abilities and personality profiles.

"Of course, timing is everything. Free enterprise is truly a great thing, but it can get ugly when new disruptive technology causes panic and pain to the market and the competition. This is something that we all will have to manage.

"Each one of you has already played an important role in the development of this product," described Manta. "The business plan will show that all you have to do is put in the key, start the engine, and put your foot on the gas."

Manta paused and took a deep breath. He noticed that he still hadn't seen much movement from the group. They all seemed frozen, or shocked, with the information. They were like stunned deer.

TWENTY-YEAR MONOPOLY

Manta changed positions to stand beside the side of the podium and lean his arm against it. This change in movement self-consciously caused the members to shift in their seats also.

He glanced out the window for a brief moment and then continued.

"As you are aware, the governments of the world allow for inventors to maintain a monopoly on their new technology, typically for twenty years. This monopoly will be granted to us in exchange for our disclosure of how the invention works. Of course, we will maintain our trade secrets.

"I am a firm believer in the justification of issuing patents. In fact, other than the laws of life, liberty, and justice for all, in my opinion the ability to patent and own technology is probably the greatest law on the books. The investors are rewarded for their risk, and the public benefits from new affordable products and services.

"The Business Plan shows how we will optimize the position of our monopoly.

"There will be greater standardization, less waste, more efficiency, and less duplication of efforts. After the twenty years is over, there will be a whole new industry born that will be driven on the competition of providing the customer with better service for less.

"The idea of a monopoly is generally not well-perceived by the public. But if you look at the history of some of the biggest monopolies, there were silver linings. For example, during AT&T's monopoly of the American telephone service, they were able to develop and standardize the protocols for the complex switching requirements and were able to wire the entire country at an astonishing rate. These models of standardization are also what made the Internet possible.

"The public may perceive that a monopoly is unfair and only allows the rich to get richer," Manta continued. "But this perception is flat-out wrong. The provision for an inventor to have rights to his property for twenty years is the best example of the redistribution of wealth possible. In our case, based on your proven management styles, we will eventually be publically traded, allowing for all levels of society to have a share of ownership. There will be fair and equitable vendor and service provider agreements. Again, in the spirit of private enterprise, it will be our goal to be the greatest example of the distribution of wealth in the history of mankind.

"I expect that quality of life will be a prime reason to work for this corporation, and that applies to everyone, from the top CEO down to the person sweeping the parking lot. There should be a great satisfaction at seeing the success and growth of our venture. Hard work should be rewarded, and slackers should get the nudge.

"The Business Plan is only a model that I think could work. It is up to you to decide which way to go and when and how to act. It is up to you to decide how long to keep the invention private and when to make it publicly owned. You have been chosen for this work, and I have full trust in you to make the right decisions.

"As stated in the plan, although we will have a substantial manufacturing base, a large revenue stream will be associated with the licensing agreements of our technologies."

Manta glanced at the clock and saw they were basically out of time. He brought his attention back to the group and continued briefly.

"Something that you will notice in the Business Plan is that I am not an officer in the company. I will remain on the board of directors and only for as long as I am needed. But, there is other important work that I must give my attention to. I look forward to sharing this information with you, at another time.

"I actually just noticed we are practically out of time. I apologize for continuing on for so long today. There was so much information to pack into this eight-hour meeting."

Bill thought quickly, cleared his throat, and then spoke up. "I don't know about the rest of you, but I can personally say that I would be willing to adjust my schedule tonight in order to stay here for a few more hours. I would need a break to call my family to let them know I will be home later than expected, though."

Bill addressed the other group members without delay to see how they felt and if anyone had a conflict of interest. What Manta had been presenting to the group was of such gripping subject matter that none of the group members were ready to adjourn the third meeting. The consensus was to take a break so everyone could call their families, Manta included, to let them know they would be home later tonight.

CHAPTER 10

PREPARATION FOR THE NEXT GREATEST CALAMITY—MAGNETIC POLE REVERSAL

After checking in with their families, they all returned. It had been a long day, but they were eager to hear Manta's disclosure to the end. Manta again stood at the front of the room and addressed the group.

"Earlier, I described how Benjamin Franklin had a fifty-fifty chance of getting it right, when he guessed the direction of the flow of electrical 'fluid.' After his identification of the positive and negative polarity, still in use today, he was a hundred and eighty degrees off from the truth, and, to this day, electrical current flows is the opposite direction of the actual electron flow—the fluid that Benjamin Franklin harnessed with his static electricity experiments.

"How could this happen? Why didn't the scientific community adjust its thinking? It happened because there were interesting rivalries between different branches of electrical experimentation in the new field of electrical phenomena during the mid-1850s.

"Earlier, in about 1820, Alessandro Volta of Italy built the first battery. By 1830, Sir Humphry Davy used Volta's battery to discover electrochemical processes to isolate elements. And in Humphry's greatest accomplishment, he mentored Michael Faraday, who succeeded him in all respects by inventing demonstrable models of the electric motor and the electric generator, and also laying the groundwork for the disciplines of magnetic induction and electrochemistry.

"Interestingly enough, Faraday was an informally educated practical thinker. He became an assistant to Sir Humphry Davy while he was apprenticed as a bookbinder. He presented Davy with a bound volume of his handwritten notes from Davy's public lectures. Davy recognized Faraday's talents and immediately hired him as a lab assistant.

"Even though all of this advancement was occurring during this period in time, the scientific leaders were struggling to understand the mechanisms for the observed electrical phenomena. Faraday, known as 'The Great Experimenter,' constructed his inventions from his practical observations. In other words, he did not know exactly what was causing the forces that made his electrical induction devices work.

"This backdrop provided fertile grounds for a competitive environment between staunch believers in electrostatics (experts in the field of static electricity), advocates of electrical induction, and experts in electrochemistry.

"In the end, the advocates of induction won, and the world adopted their model for the direction of the electrical current. One particular heavyweight in the field of electrical induction during the latter 1800s was James Clerk Maxwell. Maxwell's equations are touted today as the most significant equations in the development of science. But it is very important to understand that Maxwell's equations tell us nothing about the cause of the magnetic induction. His equations only describe the magnitude and direction based on his definition of the direction of the electrical current and the magnitude of the charge. Again, the current in Maxwell's equation is opposite to the actual flow of the electrons.

"On a side note, it is no coincidence that Einstein's work on relativity borrowed heavily from the Maxwell model, both scientifically and politically. The great thing about Maxwell's equations is that you can explain the direction and magnitude of certain phenomena, but you do not have to explain what causes it. This was a very helpful model to get over the ether dilemma. A thorough study of Einstein's success with relativity will show that putting the ether issue out of its misery was really his greatest accomplishment.

"So what does this mean for us today?" Manta asked the group.

RIGHT-HAND RULE

Manta flipped to the next page of his notes on the podium and continued.

"Part of modern conventional electrical practice (not theory), and Maxwell's equation, is the right-hand rule. What this means is that when a direct electrical current (DC) is sent through a conductor (wire), the electrical current in the conductor causes an induced field to occur in the direction that your figures point, when you point your right thumb in the direction of the current," Manta explained as he visually showed the group what he meant by using his hands.

"Please understand that this induced field I am talking about is not some imaginary thing. It is very real. For example, if you were to send a reasonable amount of electrical current through a wire, say fifty amps, and at the same time you were holding a small piece of magnetic material, such as iron or steel, about four inches from the wire, the induced field would launch that piece of metal across the room. The direction the metal would be launched is in accordance with the right-hand rule, you see.

"Now, again, the electron flow in the wire is traveling in the opposite direction as the electric current, so using a left-hand rule would more accurately depict the direction of the field relative to electron flow and its affect on the piece of metal. But, in the end, the common acceptance of the standard current flow being opposite to electron flow did not prevent science and industry from designing and constructing electric motors and generators to power our cities.

"This application of induced fields has great significance in our everyday modern life. Power plants have generators that spin electromagnets that induce electric fields producing twenty thousand volts, or more. This kind of power would vaporize you in a microsecond.

"The operation of all power devices, such as electric motors and generators, and some electronic devices, such as inductors in an electronic radio circuit or power supply, are all designed with the right-hand rule. The direction of the induced field is pointed in the correct direction in order for these systems to operate correctly." Manta gestured with his hands again to emphasis what he was explaining.

"Why does the induced field point in the direction of the right-hand rule? Nobody knows. That is, nobody *knew*. The FNU Wikappendix reveals the mechanisms of electrical induction based on our discovery of the Spirogrid and the ethertron theory.

"The discovery of the Spirogrid reveals to us that there is a left-hand bias in the universe," Manta explained as he showed what a left-hand bias would look like with an illustration of a spirograph on the whiteboard. "Notice how there is a constant turning to the left as the spirograph is formed.

"Please note that I am speaking as though we are using a true left-hand rule, which is based on the correct direction of the induced electrons in an electrical circuit.

"Now that you have been introduced to the ethertron, the corporeal component of the Spirogrid, examination shows that we, our whole galaxy, are constantly turning to the left relative to the corporeal motion of the ethertron, even though we do not feel it. It is similar to the fact that we are actually standing horizontally on the side of the curved earth and spinning at 1,670 km/hr, but to us everything is flat, level, and stationary.

"As described in The FNU Wikappendix, the left-hand bias of the corporeal motion of the Spirogrid is the cause of the direction of the induced field.

"However," Manta paused before continuing, "something very traumatic is going to occur in the near future that will change this."

He shifted a few papers around again on the lectern as he took a few moments to gather his thoughts.

MAGNETIC POLE REVERSAL

The group members were at the point where nothing else Manta could tell them would shock them very much, even if it were shock-worthy material. No one stirred in their seats or made any type of noise during these brief moments. Manta's information was enthralling them, and they were eager to know his next reveal. A few of them, in their minds, were guessing at what he would say, but no one was sure at this moment.

Manta cleared his throat and continued.

"In the near future, at an instant of time, the Spirogridic motion of the universe will lapse from a left-hand into a true right-hand bias.

"Because of the nature of Spirogridic motion, there is a point in time and space where the spirographic motion rolls over in one continuous motion from a left-hand turning direction to a right-hand turning direction without any perceived change or effect on the existing ponderable matter," Manta explained as he again gestured with his hands in a way to show how the pattern will change.

"When this change occurs, people and animals will not notice any change, and they will not feel anything different at all. It will be as if nothing happened. There will be no change in how Earth rotates or shields itself from radiation either. Even the famous aurora borealis will seem to continue as usual.

"What will change, however, is that every human device that functions with electrically induced fields will stop working."

Manta noticed many wide eyes around the room and, still, no one moved an inch in their seats or made any type of noise whatsoever.

Manta knew this information would seem startling at first and continued, slowing the pace of the speech.

"Therefore, what this means is that electric motors will run backward, if they even will be able to run at all. Generators will stop generating. Electronic power supplies and circuitry components that rely on magnetic induction will stop functioning as well.

"Now listen carefully. I am not concerned about electric power, motors, generators, or electronic devices, even though these are all very important things we use almost every day. I am not concerned because these devices can be fixed once the problem is discovered. What I *am* concerned about is magnetic memory media.

"It is very possible that every hard drive in the world may be wiped clean. I have been investigating the effects on flash memory, but, as of now, the jury is still out on whether it would be recoverable or not.

"One thing I can tell you for sure is that the Internet will immediately go down. Whether or not it can be recovered, and how much time it will take, are complete unknowns.

"I have included a contingency plan to address this as part of the Business Plan, so, whenever this magnetic pole reversal occurs, it will not put an end to the corporation I have proposed to all of you.

"With that, ladies and gentlemen, I have completed my disclosure." Manta stood motionless for a moment. "What would you all like to do?" Manta then asked the group, as he panned his eyes into the eyes of each of the twelve.

There was a silence in the room as all twelve immediately thumbed through the Business Plan to find the section on magnetic pole reversal. One main question resonated through all their minds.

"When is this pole reversal set to occur?" asked Walter, voicing the concerns of all.

"That I am unable to tell you," said Manta. "I don't know why I can't give you an exact answer on this, but I am completely blank. I have thought a lot about it, and it is possible that maybe it can't be known for sure because there is some randomness to the exact time it will take place. For example, if you spin a top and let it go spinning onto the floor, can you tell exactly when it will topple over? No, but what you do know is that it is bound to topple over at some point. With the span of time associated with magnetic pole reversal, it could happen tomorrow, or it could happen in fifty years. All I know is that we have to get the contingency plan in order as soon as possible, because it is bound to happen at some point," said Manta.

"When you read the pages of the contingency plan, which I believe you are all on now, you will see that it does not involve a broad announcement of pending disaster. There will be a systematic approach to making the public aware of these issues and concerns, and then the market will take over to supply the products that address the problem.

"Basically, data recording media will be advertised as 'pole-reversal safe.' The important part will be on the manufacturing side. Namely, memory reading devices will have to be designed to operate correctly or be able to be easily switched over when the magnetic induction switches from the *false* right-hand rule, to the *true* right-hand rule," explained Manta. "Of course, our company already has the IP in place and we will be involved in the licensing and distribution of these products."

"OK, I think we all have the gist of it," said Bill. "We all know what we have to do now. We have to vote on whether we are in or out on Miles's new corporation."

Bill paused briefly, then continued. "I can only speak for myself, but I am excited to be part of this effort. The transition will be incredibly smooth to move on from my company—obviously no sheer coincidence after learning about Mr. Manta's activities." Bill looked around at the others to gauge their interest in Manta's business proposal.

"It's been a crash course, but we have all had plenty of time to search this over and sleep on these new ideas Miles has presented to us. We could have a vote now to find out where we stand, and then go from there. What do you think?" Bill asked the group.

"I make a motion that we take a vote to determine if we are all in," suggested Malik.

"I second the motion," spoke up Katerina.

"OK, we have a motion on the floor, and a second," said Bill. "Who is in favor of an 'all in' effort for the energy corporation described in the Business Plan set before us authored by Mr. Miles Manta?"

The vote was an immediate twelve ayes, which caused instantaneous chills throughout the room. The demeanor of the group remained professional, but there was definitely an air of excitement floating about.

"What about you Miles, what is going to be your next crusade?" asked Bill.

Manta cleared his throat and took the floor again.

"There will be a lot more for us to talk about, including the unfulfilled promises I made to some of you to answer specific questions about the future. For now, we all understand that there is an immediate need to begin this momentous enterprise of solving the world's energy and environmental crisis, and you are to be the key players. As for me and my next crusade, now that I have revealed myself, I cannot sit back and do nothing about the next potential Chebala disaster. As you would guess, there will be efforts to weaponize this new technology to gain military superiority. My role in how all of this transpires will depend on the form of my adversary. So, other than that, let's get started…"

The end

HANDOUT A - THE MAKING OF AN EQUATION, NEWTON'S 2ND LAW, AND $E = MC^2$ REVISITED

In a very large way, the technical advances that we see manifest in our world today exist because of our ability to form mathematical equations that model the forces and behavior of matter. Mathematical expressions are often the language of new breakthroughs in science and technology.

How are these equations made? Some are made by pure intuition, others by the stroke of calculus.

First, I would like to start off with some examples of pure intuition.

EXAMPLE OF EQUATIONS MADE BY INTUITION

OHM'S LAW

All electronic devices are designed with the help of an equation called Ohm's Law. Ohm's Law is as follows:

$$V = IR.$$

Where V = voltage (units of "volts"), I = electrical current (units of "amperage"), and R = resistance (units of "ohms")

To me this equation is obvious, especially if we rewrite it in the form:

$$I = \frac{V}{R}.$$

It is obvious because we know from the earliest days of Ben Franklin's experiments that electrical current is analogous to the flow of an incompressible fluid in a pipe. From Alassandro Volta's experiments and development of the first volt meter, we know voltage is analogous to the pressure that pushes fluid through a pipe. Therefore, we can intuitively form the following expression:

$$I \propto V,$$

Which says "the flow of current is 'directly' proportional to the voltage". Or, in other words, "when the voltage goes up, the current goes up".

Now we know that pipes create "resistance" to flow of a fluid, in other words, it is easier to push water through a large pipe (say 2.5 centimeters in diameter), then through a small diameter pipe that is half the diameter of a plastic drinking straw. If you don't believe me, fill your mouth with water and try to blow the water through a small straw and compare it to a pipe the same length but ten times the diameter.

Intuitively, we can say "the flow is 'inversely' proportional to the amount of resistance", or in other words "the higher the resistance the lower the flow". Installing the relationship into our prior expression, the relationship can be expressed mathematically as follows:

$$I \propto \frac{V}{R} \ .$$

In this example of developing a mathematical relationship there are no other factors. Amazing, this is all there is! Since there are no other factors influencing the flow of electrical current, we can get rid of the "proportional" symbol, α, and substitute the all-powerful equal sign, $=$, as follows:

$$I = \frac{V}{R} \quad , \text{ which can be written as the more familiar,}$$
$$V = IR \ .$$

Again, please understand the significance of this equation. A modern computer chip contains a massively complex electrical circuit with hundreds of millions of transistors and resistors. Each one of these infinitesimally small components operates exactly according to this equation, at the precision of the atomic scale.

Another example of an intuitive equation is the exact expression of what we define as power, $P = VI$, where P = power (watts), V = volts, and I = current (amps). This expression can be derived intuitively because we can reason that power is directly proportion to the value of the two factors, voltage and current. The expression is exact, or compete, because like Ohm's Law, there are no other factors to consider.

THE EQUATION FOR ELECTRICAL RESISTANCE, R

Now let's use our ability to make equations based on intuition to address a problem related to determining the "resistance", R, of electrical current flow in a pipe, or a conductor. Remember, fluid flow in a pipe and electrical current flow in a conductor are analogous.

As we already discussed, the flow in the pipe is more difficult when the pipe is smaller in diameter, therefore we know that the resistance in the pipe is "inversely" proportional to the cross sectional area of the pipe, expressed as follows:

$$R \propto \frac{1}{A}$$

Where R is "resistance" , and "A" is the cross-sectional area of the pipe, or conductor. Now, intuitively, we know that the longer the pipe or conductor is, the more resistance it will have to the flow of fluid in the pipe. This is a "direct" relationship. Therefore, we include this factor into our expression as follows:

$$R \propto \frac{L}{A}$$

Where "L" is the length of the pipe or conductor.

But this equation is not complete, because we know intuitively, that the resistance to flow is also a function of the fluid density, or in the case of electrical flow, the "conductivity" or "resistivity" of the conductor. For example, copper has much more "conductivity" than steel. Therefore, we can intuitively say that that "the flow of electrical current is 'directly' proportional to the conductivity of the metal", and this can be factored into our equation as follows:

$$R \propto \frac{\sigma L}{A}$$

Where "σ", Greek letter sigma, is used to express the value of "conductivity" of a conductor, typically a metal. Please note that we could have used the expression of "resistivity", ρ, Greek letter rho, but it would have to be placed in the denominator, because the flow of current would be "inversely" proportional to the resistivity of the conductor. Conductivity is a unique physical property of each conductor material, and according to modern science, we define the value of conductivity as the inverse of the value of resistivity (units of "ohm-cm"). Scientists pick and choose which expression they prefer to use.

Unlike the intuitive derivation of Ohm's Law and, where the derivation ends with an absolute final expression, the equation for resistance, R, has room for additional factors. We could add other factors into our equation, such as temperature, pressure, that could change the value of conductivity, σ. However, for basic work in the field of electronics, designers use the equation in this form:

$$R = \frac{\sigma L}{A}$$

Please understand that what makes the equations above intuitive is that we have a good analogous understanding of the flow of electricity compared to the flow of water. A lot of credit must go to the pioneers of science, like Isaac Newton, Michael Faraday, Ben Franklin, Alessandro Volta, Louis Ampere, and George Simon Ohm. They did not have the benefit of understanding the "units" of measure that we take for granted, they had to invent them. Again, we have the benefit of standing on the shoulders of these giants.

MAXWELL'S EQUATIONS – EXAMPLE OF BLIND INTUITION

An example of what could be considered the highest level of intuitive equations could be what are referred to as "Maxwell's Equations".

Let's look at the expression of what is called "Gauss's Law" for electric fields. In an overly simplified manner, the equation is as follows:

$$E = \frac{\rho}{\varepsilon_0} \ .$$

Where E = strength of the electric field (Newtons per Coulomb), ρ = charge density (Coulombs per cubic meter), and ε_0 = electrical permittivity of free space.

Intuitively, the expression is saying "the strength of the electric field is 'directly' proportional to the charge density of the source of the charge and 'inversely' proportional to some dissipating factor of the surrounding medium called 'electric permittivity'".

Sadly, with Maxwell's Equations, the intuition ends at an early stage, for a combination of reasons, including scientific and philosophical.

The scientific problems stem from the fact that electric fields are three dimensional, and are influenced by motion. For example, when you move a magnet past a wire that is part of a continuous circuit, the "field" of the magnet induces a current in the circuit that fights back against the motion. Also, the "field" of a natural magnet, induction coil, or electromagnet, has directional properties. Therefore, the true expression for Gauss's Law, is as follows:

$$\vec{\nabla} \, \alpha \, \vec{E} = \frac{\rho}{\varepsilon_0} \, .$$

The arrow symbols and the little circle are very complex mathematical operators that explain the relationship as a "vector", "differential", and "dot product", with non-intuitive meanings of "divergence", "curl", and "gradient". Very few people can master these concepts, and my hat goes off to those who can comprehend this high level math. The philosophical problem stems from the fact that with Maxwell's Equations, no direct definition of what the electric field is has been given. The text books only state that it is "said to exist". Unlike the intuitive examples provided above for ohm's law, where we can envision water flow in a pipe, with magnetic induction there is a huge difficulty in understanding by way of analogy, the real meaning of the factors. It becomes a matter of philosophy because it has resided in the realm of opinion, speculation, dogma, and faith.

When you search for the meaning of Maxwell's Equations there seems to be no end to the layers of complexity associated with explanations of the many electromagnetic phenomena. Wading through the displays of double and triple integration, and the attempt to discern real value to the myriad of esoteric equations, can be a disheartening experience. A search of the applications of Maxwell's Equations shows that they don't always work, and only more advanced "quantum" models are claimed to be accurate, requiring more faith. Sometimes you can surmise that the "quantum" explanations are only fanciful words to say that we have no clue what is really going on. Maxwell's Equations, and the lack of a mechanistic model to provide a physically-based understanding of electromagnetic phenomena, provided Einstein and the barraters of Newton's ether theory, a footing to gain supremacy. The acceptances of Maxwell's equations, without the mechanistic model, were exploited by an elite group within the scientific community to obtain power and notoriety, by conjuring up advanced theories without having to prove anything. The basis of their theories was that the ether is "superfluous", so anything goes, and it will go the way we say. Now, with the understanding of the Fundamental Nature of the Universe, and the spirogridic motion of all matter, there will be a tremendous advance in our understanding of the interactions of matter and energy, including electrical induction. Now, let's examine the development of non-intuitive equations associated with E = mc².

NEWTON'S 2ND LAW AND E = MC² REVISITED
(CALCULUS BACKGROUND RECOMMENDED)

Most of what we know about the behavior of forces and motions comes from the work of Isaac Newton (1643–1727). It was Newton who successfully defined, with mathematical equations, the relationship between what is mass, momentum, and force. His work gave us the ability to add, subtract, multiply, and divide the effects of force and motion.

But there was something more that came out of his work, maybe more than he ever realized. It was the ability of the powerful tools of calculus, namely, derivation and integration, to guide us in our discovery of unknown forms of force, motion, and energy.

The powerful tools of derivation and integration have allowed us to discover and understand things that are beyond human intuition. One of the most famous examples would be the equation that brought Albert Einstein fame, $E = mc^2$.

Credit must also go to another principal player in the discovery of calculus, Gottfried Leibniz (1646–1716). It is his systems of notation that are used today to show the operators of derivation and integration.

But, it is Newton who gets credit for defining what we know as mass, force (or weight), and momentum. From these basic quantities we derive all the known values of force and energy.

When Newton started his work there was only a rudimentary understanding of the concept of mass. Galileo, Kepler, and Hooke had developed theories related to the mass of an object, but the concept was not fully understood.

What is paramount about the concept of "mass" is that Newton used it to define "momentum." This is the pivotal breakthrough that has allowed great progress and discoveries.

Momentum is simply the mass of an object multiplied by its velocity, or,

$$\text{Momentum} = M = \text{mass (kilograms)} \times \text{velocity (meters per second)} = mv$$

So, why is the concept of momentum important? He used momentum to define the second law of motion, which states:

"The rate of change of momentum of a body is equal to the resultant force acting on the body and is in the same direction."

Like velocity is equal to the change in distance per unit of time, force is equal to the change in momentum per unit time. Newton was exactly correct; it works. The mathematical description of this concept is as follows:

$$\text{Force} = F = \frac{\text{Change in Momentum}}{\text{Change in Time}}$$

$$F = \frac{(\text{mass x Velocity 1}) - (\text{mass x Velocity 2})}{(\text{time 1 - time 2})}$$

Or, as scientists and engineers express,

$$F = \frac{\Delta mv}{\Delta t}, \text{ where m = mass, v = velocity, } \Delta = \text{Delta = Change or Difference.}$$

Or, using modern Leibniz notation for Calculus,

$$F = \frac{d\,mv}{d\,t} \text{ where m = mass, v = velocity}$$

where, $\frac{d\,mv}{d\,t}$, says "the instantaneous value of $\frac{d\,mv}{d\,t}$ as the value of dt approaches zero."

One reason that the use of calculus is extremely important is that without it, the expression for force with Δt in the denominator, $F = \frac{\Delta mv}{\Delta t}$, cannot be solved for the "instantaneous" value.

Without calculus, instantaneous velocity would require a value of zero in the denominator, $\Delta t = 0$, which of course, is not allowed; it is undefined.

This is similar to the "Area under the Curve" problem that perplexed mathematicians for decades before the discovery of calculus. It may seem subtle, but it is not; the rules of calculus allow us to overcome this serious roadblock.

Anybody who has studied engineering or physics will recognize the equation drawn from Newton's second law of motion:

$$F = \frac{d\,mv}{dt}.$$

Because the value of mass is constant, the equation is more often shown as:

$$F = m\frac{dv}{dt}, \text{ where } \frac{dv}{dt} = \text{acceleration} = \text{a. Therefore,}$$

$$F = Force = ma\,.$$

And this equation is the most famous in all science.

This equation is what has defined the motion of the planets, and forms the basis of all materials standards and specifications.

This equation is what tells you how much you weigh. For example, you might weigh 150 pounds (force). In the more sensible metric system, your "weight" is not specified in terms of force, but rather in mass, where a 150 pounds force is produced from a 68 kilograms (mass), in Earth's gravity. The units for mass in the American system are "slugs," so in the American system of measurement your mass is 150 pounds divided by Earth's acceleration of 32.2 ft/sec,[2] therefore 150 /32.2 = 4.7 slugs (mass). Slugs? Another example why America needs to adopt the metric system as soon as possible.

THE PRODUCT RULE OF CALCULUS—THE OTHER HALF OF NEWTON'S 2ND LAW

With the invention of calculus came several specific "rules" that can be used to solve problems of integration and differentiation. One such powerful tool is the "**product rule**" of calculus, also called "Leibniz's Law." As with other procedures invented with the mathematics of calculus, the "product rule" is accurate in the same manner as addition, subtraction, multiplication, and division.

The **product rule** described in mathematical terms is as follows:

$$F = \frac{d\,(xy)}{dt} = x\frac{dy}{dt} + y\frac{dx}{dt}\,.$$

As described above, Newton's discovery of the 2nd Law of Motion is:

$$F = \frac{d\,(mv)}{d\,t}.$$

Remember before when we introduced F = ma, above, and we said the mass is a "constant" and does not change. With mass being a constant, this allowed us to move the value of mass outside of the differential (mass does not change with respect to time). But, what if the "mass" does change? This was Einstein's claim. Then it is not a constant and we cannot move it out in front of the equation.

If mass is not constant, it should be noticed that the true solution to Newton's 2nd Law of Motion equation, according to the product rule of calculus, is as follows:

$$F = \frac{d\,(mv)}{d\,t} = m\frac{dv}{dt} + v\frac{dm}{dt}.$$

In plain language this formula states that Force is equal to the quantity of mass multiplied times acceleration (F = ma, which we use on a daily basis) <u>plus</u> the velocity multiplied times the change in mass per change in time.

Newton presumably cancelled out the second part of the equation, believing that $\frac{dm}{dt}$, the change in mass over the change in time, is zero. Again, this would make sense, because why would the mass change? Mass is matter, and matter is conserved.

However, during Einstein's period in the late 1800s and early 1900s, people where experimenting with concepts of sub-atomic particles, including electrons, and the elusive ether. Could the mass of the atomic particles change? Could this second part of the equation, $\frac{dm}{dt}$, be used to derive E = mc²?

This expression of Newton's second law in terms of the product rule is nothing new. Of course Newton himself would have expressed the derivative $\frac{d\,(mv)}{d\,t}$ in this manner (in his own style of notation). But, let's put ourselves back in time to Newton's time.

Newton lived between the years 1643–1727. This is before the discovery of the atom, before the discovery of the elements, electricity, chemistry, or the periodic table. Newton's frame of mind about the universe was primarily focused on the macroscopic realm, the motion of physical objects such as the earth, moon, and planets.

We don't know how far he took it, but the expression of $v\frac{dm}{dt}$, where $\frac{dm}{dt}$ says "change in mass divided by the change in time," apparently was not taken very far by Newton.

We know that Newton was an advocate of what he called the "luminiferous ether." One would think that he would try to relate the ether to the expression $\frac{dm}{dt}$.

ALONG COMES EINSTEIN

It was the evaluation of the relationship between mass and energy where a "Y" in the road occurred in the modern understanding in the nature of the physical universe. But, enter another factor, light.

Why do we have to add light into the fray? Because there was a controversy about the nature of light and—forget about the science—controversy is good sport.

The controversial issue of the ether was coming to a head. The ether was supposed to be the "medium" through which the "waves" of light travelled. But the Michelson-Morley experiment was showing there was no change in the speed of light relative to the motion of the earth. The mechanistic explanations of the ether had been picked apart by the learned men of science. The proponents of the ether simply could not figure out an ether model that worked. Probably the last formal dissertation on the subject of the ether was given as the Nobel Lecture in 1902, the same year Hendrik Lorentz received the Nobel Prize, and was titled "The Theory of Electrons and the Propagation of Light," where the theory describes "the physical world as consisting of three separate things, composed of three types of building material: first ordinary tangible or ponderable matter, second electrons, and third ether."

It was not long after this when Lorentz issued his famous statement, "I am at the end of my Latin," when he gave up trying to defend the ether model.

Einstein offered an alternative to a tangible ether. His alternative included the wave-particle "duality" of light. The dual nature of light he proposed is that light behaves as a wave in all respects of the well-established laws of wave propagation and interference that apply to normal sound waves, but, in the absence of a medium, light propagates as a particle, called the "photon." Therefore, his theory did not require an ether medium, or, it was "superfluous".

You can conduct your own Internet search to find examples showing the derivation of E = mc² offered by Einstein. It generally describes a "thought experiment" where there is a box in space, and a photon is launched from left to right imparting its momentum to the side of the box. The conservation of momentum is applied to the "mass-less" photon, because apparently under certain circumstances you can assign mass to the mass-less photon.

This derivation by Einstein is the cornerstone of the gospel of modern day particle physics.

From the equation E = mc², scientists believe that the energy of one kilogram of sugar (or any other "mass") to be equivalent to 21.48 megatons of TNT. This energy is attributed to the "intrinsic" energy of the mass, and that mass *is* energy, and vice versa.

You will find when you research the derivation of E = mc² that there are solutions ranging from one page, to claims that it would take an entire book to show the derivation. The more you study the laws of physics the more you should see how there are different approaches to arrive at the same conclusion. This is actually to be expected of something that is verifiably true, and is sort of proof. It does not matter that there are more complex derivations; the simple derivations should be adequate for communicating the theory.

The real problems lie with the folks that believe you must solve this problem with "quantum mechanics" and "relativistic" models exclusive to and only understood by an elite class of the priesthood. In my opinion, if it can't be explained to a person of average intelligence and education, then it begins to weigh more heavily on "faith" rather than true science. Be wary of people who want to make these concepts more complicated that they have to be.

E = MC² REVISITED

In the derivation of the Complete New Energy Equation, see Figures 5 through 10 of Chapter 7. We used the process of <u>integration</u> to determine the area under the curve that occurs when you plot momentum verses time. This derivation led us to the well-known and widely used formula for "kinetic energy," $E_k = \frac{1}{2}mv^2$.

In addition, the process of integration, through the required "constant of integration," led us to the understanding of "transformic" energy, E_t, a form of intrinsic energy of mass. Furthermore, the process of integration produced a component of "potential" energy, E_p, which has known and possibly unknown forms.

From the integration process and the model of spirographic motion of the universe, we derived the Complete New Energy Equation, as follows:

$$E_{Total} = E_t + E_t = [\tfrac{1}{2}mv^2 + M_{to}v + e_{to}] + [\tfrac{1}{2}Iw^2 + M_{Io}w + e_{io}]$$

Arranging these into the three (3) fundamental categories of energy, we have:

$$\textbf{Total Energy Equation} = E_{Total} = \left[\tfrac{1}{2}mv^2 + \tfrac{1}{2}Iw^2\right] + \left[M_{to}v + M_{Io}w\right] + \left[e_{to} + e_{io}\right]$$

| Total Kinetic Energy | Total Transformic Energy | Total Potential Energy |

where, m = mass, v = translational velocity, I = moment of inertia of mass, w = angular velocity, and:
M_{to} = Translational, or linear, momentum of mass due to spirographic motion
M_{Io} = Angular, or rotational, momentum of mass due to spirographic motion, and:
e_{to} = Potential energy, source unknown (energy, work, power, mgh, pressure x volume?)
e_{io} = Potential energy, source unknown (energy, work, power, mgh, pressure x volume?)

Let's now look at what can be done with the <u>derivation</u> of momentum from Newton's 2nd Law and the product rule. We know:

$$F = \frac{d\,(mv)}{dt} = m\frac{dv}{dt} + v\frac{dm}{dt}.$$

For extremely small, nearly mass-less objects travelling near the speed of light, let's assume that they do not slow down or accelerate, therefore, $m\frac{dv}{dt} = 0$, or at least approaches zero, therefore:

$$\text{Force} = v\frac{dm}{dt}.$$

It is a fact of science that Energy (work energy) is equal to Force multiplied times distance, or:

$$\text{Energy} = \text{Force } \Delta x = F \, \Delta x \text{ , Where } \Delta x = \text{change in distance.}$$

Or, as stated using calculus, that Energy is the area under the curve when you plot Force verses Distance, therefore:

$$\text{Energy} = \int F \, dx.$$

Substituting in the value of F:

$$\text{Energy} = \int v \, \frac{dm}{dt} \, dx,$$

$$\text{Energy} = v \int \frac{dm}{dt} \, dx = v \, \frac{dm}{dt} \, x \, .$$

So, now we have an expression that says the Energy is equal to the velocity (v) multiplied by the change in mass divided by time $\left(\frac{dm}{dt}\right)$, multiplied by a distance value (x).

According to the rules of algebra, this equation can be written:

$$\text{Energy} = dm \, v \, \frac{x}{dt}.$$

Notice in the above equation that $\frac{x}{dt}$ is an expression for velocity. Therefore, like the example provided in Figure 9 of the main story, where we played with the possibility of substituting values for the speed of light into the transformic energy equations, we can suppose that the value of "v" is the velocity of the motion of the Spirogridic pattern travelling at the speed of light (v = c), and the mass was translating, moving through space, at the speed of light ($\frac{x}{dt}$ = c) . Then the above equation becomes:

$$\text{Energy} = m \, v \, \frac{x}{dt} = mcc = mc^2.$$

But, as described in Figure 9, there are a number of problems with the popular opinions about his equation. The energy of the mass is not intrinsic, as most believe, but is associated with the kinetic energy of the motion of the mass. If such a translational velocity were possible, the $E = mc^2$ expression would only be a portion of the total energy of the mass. What proof do we have that $v_{grid} = c$? Maybe it's higher. Also, it brings up questions about possible limits to the value of translational velocity, and the true relationship of translational verses corporeal motion.

So, let's look one more time at the <u>derivation</u> of momentum from Newton's 2nd Law and the product rule:

$$F = \frac{d\,(mv)}{dt} = m\frac{dv}{dt} + v\frac{dm}{dt}.$$

Several possibilities are suggested. First, mass is conserved, and it cannot be created nor destroyed, therefore the value of $\frac{dm}{dt}$ is zero. A second possibility is that the value of $\frac{dm}{dt}$ is related to a change in mass caused by mass being added or taken away from the system.

A third possibility is that it is related to the change in the geometric positioning of mass, and that Newton's second law itself has to be rewritten in terms of I, the moment of Inertia. For example, the point mass has a moment of inertia. Examples of the effort will be provided in future discussions contained in the FNU Wikappendix.

A better option is, rather than trying to make $E = mc^2$ fit some preconceived construction, we should abandon it completely and focus our energy on applications of the New Complete Energy Equation, as described in Figures 5 through 10 of Chapter 7, and Handout B.

Handout B – Summary of Complete New Energy Equation

Equation No. 1 - Distance Equation

Modern science has provided us with a mathematical equation to determine the exact distance traveled over a period of time at a given velocity and acceleration. As follows:

$$\text{Distance} = d = \frac{1}{2}at^2 + v_0 t + d_0$$

Where, d = final distance, a = acceleration, and v_0 = constant velocity before acceleration, and d_0 = distance form origin before acceleration and v_0 occurred. See detailed derivation in Figure 5.

Equation No. 2 - Energy Equation for a mass moving in a line, or in "translational" motion

With the same method used to derive the distance equation, we use Isaac Newton's equation for momentum ($M = mv$), to calculate the exact "translational" energy of a moving mass. See Figure 6 for detailed derivation.

$$\text{Total Translational Energy} = E_t = \frac{1}{2}mv^2 + M_0 v + e_0$$

Where, E_t = total translational energy of mass following a "line", m = mass, and v = present translational velocity of mass, and M_0 = momentum of mass before added velocity, e_0 = unknown potential energy of mass before v.

Equation No. 3 – Energy Equation for a rotating mass, with angular velocity, w

With the same method used to derive the distance equation, we use the equation for angular momentum ($M = Iw$), to calculated the exact "rotational" energy of a rotating mass. See Figure 8 for detailed derivation.

$$\text{Total Rotational Energy} = E_I = \frac{1}{2}Iw^2 + M_{Io}w + e_{Io}$$

Where, I = moment of Inertia for mass (function of mass and geometry, units of kg-meter²), and w= present angular velocity of mass, and M_{Io} = angular momentum of mass before added angular velocity, e_{Io} = unknown potential energy of mass before w.

Equation No. 4 – The New Total Energy Equation for all mass in Universe

As far as we know, all mass in the universe is moving, both translating and rotating. Therefore, the total energy of any mass is the summation of Equation No.2 and No. 3. See Figure 9 for details.

$$\text{Total Energy Equation} = E_{Total} = E_t + E_I = [\frac{1}{2}mv^2 + M_{to}v + e_{to}] + [\frac{1}{2}Iw^2 + M_{Io}w + e_{Io}]$$

Arranging these into the three (3) fundamental categories of energy, we have:

$$\text{Total Energy Equation} = E_{Total} = \left[\frac{1}{2}mv^2 + \frac{1}{2}Iw^2\right] + [M_{to}v + M_{Io}w] + [e_{to} + e_{io}]$$

Total Kinetic Energy, E_K	Total Transformic Energy, E_{Tr}	Total Potential Energy, E_P

Where, m = mass , v = translational velocity, I = moment of inertia of mass, w = angular velocity, and,

M_{to}= Translational, or linear, momentum of mass due to spirographic motion = mv_{grid}

M_{Io}= Angular, or rotational, momentum of mass due to spirographic motion = $I_{Io}w_{Io}$, and,

e_{to} = Potential energy, source unknown (energy, work, power, mgh, pressure x volume?)

e_{io} = Potential energy, source unknown (energy, work, power, mgh, pressure x volume?)

NOTE: The expression for Total Transformic Energy, E_{Tr}, makes the expression E=mc² obsolete, and can be applied to conventional chemical and thermal reactions, see Figure 10.

INTRODUCTION TO THE FNU WIKAPPENDIX

The appendix following this introduction is an incomplete outline, on purpose. Furthermore, this appendix will be part of an ongoing work that includes an online FNU Wikappendix, blog, and newsletter.

The purpose of this Wikappendix is to invite participation by all parties interested in advancing the ideas and concepts introduced by Miles Manta. The individuals and groups that are adherents to the ideas and concepts will be encouraged to use all the means possible to provide an open forum for discussion, exchange of ideas, and evaluation of experimental data. This obviously involves the World Wide Web, including blogs and Wikipedia-style submittals and review methods.

It is a lofty expectation to think that the story of Miles Manta would provide universally accepted details describing a grand unified theory of the universe. However, like many science fiction stories of the past, an imaginative and freethinking story line can effectively plant the seeds for new ideas about energy and the physical universe.

Since the nature of the Spirogrid involves resonating frequencies, it will require more complex models, similar to the differences between AC and DC circuits. There was an order of magnitude difference in the complexity of the mathematical models understood and used by Thomas Edison for his DC circuits and Nikola Tesla for his AC circuits. There will be simple mathematical models of the Spirogrid, but who knows how far really bright individuals will develop the theories presented?

There are essential components to the discovery of new concepts, designs, and inventions. Probably the most important step is gaining valuable experience through experimentation. This essential component of new discovery includes patience and tolerance, especially associated with overcoming mistakes. How many mistakes did Thomas Edison make before he got it right? As in the case of Edison, however, the person who has made a lot of mistakes is the greatest asset. The person who understands the limitations is usually better equipped to eventually get it right.

It is my hope, as the author, that the information presented, with all its imperfections, provides motivation for new scientific understanding and accomplishments. From new scientific understanding, we can advance our standard of living with emphasis that our advancement not only protects the natural environment, but also enhances it. I am convinced that human beings and the progress they will make can synergistically coexist with the natural environment. I believe it is simply a matter of choice—do we want to? If we do, then it is simply just a matter of getting to work.

Conducting work requires energy—the word appearing on the front page of practically every newspaper every day of the week. Energy has become so essential to our existence that, God forbid, if we would run out of it, there would be desperation and calamity, both for the people and the flora and fauna of Earth. At least that is my opinion.

The primary purpose of this Wikappendix and the ensuing forum and open exchange is to look forward, not backward. All of the information can be helpful, but the most important goal of scientific discovery should be to provide for the physical needs of the world today and tomorrow, and not what happened eons ago. Let's focus on the future and how we can make the most energy-efficient progress to help benefit our planet. Does it really matter how or why the universe was formed in a Big Bang? Is it really worth our time and expense to try to understand something that happened one hundred billion years ago, or was it one thousand billion years ago, again, as if it really matters? Does anyone really believe that a theory about what happened at the beginning of the universe will change someone's perception about God?

I would like to put myself back in time to 1902, when Hendrik Lorentz received the Nobel Prize, and the Nobel Lecture titled "The Theory of Electrons and the Propagation of Light" suggested "the physical world as consisting of three separate things, composed of three types of building material: first ordinary tangible or ponderable matter, second electrons, and third ether". In honor of Lorentz and other scientists who at the turn of the nineteenth century held back from presuming notions, with upmost humility, I would like to make the following suggestion:

There are four separate things, first matter, second distance, third time, and forth nothingness (a true vacuum), from these four entities come all forms of material phenomena and physical existence.

Matter can occur in the form of ponderable mass, or imponderable mass. The ponderable mass are the elements and compounds we know, that occur in the various phases, including solids, liquids, and gases.

The imponderable mass would be in the subatomic realm, including the incompressible liquidous-like electron, and the compressible gaseous-like form of the electron that, for the sake of historical continuity, can be called the ether.

The solid phase of the electron could be the basic atomic unit of ponderable mass, which to reconcile the phenomenon of molar volume, must be of generally uniform density, but possibly of differing size, length, shape, or material profile.

The individual and unique characteristics of ponderable mass, including the known elements and compounds, are suggested to be caused by individual and unique cymatic structures that occur around a true vacuous core. This vacuous core, being true nothingness (a true vacuum), produces strong nuclear forces due to the intense "gravitational" cavity created by the resonant kinetic motion of all known matter within the universe.

Finally, a patent application has been filed on a device that functions on the concepts described in the book. The objective of this patent is clear: the quest to develop the next energy source for the world. The basic discoveries associated with the Pressure Density Cell are similar to when we discovered that diamonds were composed of pure carbon. It was only a matter of time for researchers to discover the pressure and temperature conditions to transform inexpensive chunks of carbon graphite into diamonds. Of course, success with the Pressure Density Cell will produce a whole slew of products and services, similar to other breakthroughs in technology.

You can have a part in this effort by responding to the ongoing effort with The FNU Wikappendix, or by supporting our effort of using the ideas presented in this book to produce clean and reliable energy, at www.wikappendix.com. Investor inquiries are invited. I look forward to the dialogue.

Sincerely,
Angelo Spadoni
www.wikappendix.com

NOTE: This book includes descriptions of a new branch of intellectual property. If interested in being a part of this effort, including the provision of scientific and financial assistance, please contact the author at www.angelospadoni.com.

APPENDIX

THE FUNDAMENTAL NATURE OF THE UNIVERSE (FNU) ONGOING APPENDIX

(OUTLINE VERSION 1.0)

FOR UPDATES AND

YOUTUBE VIDEO DEMONSTRATIONS SEE

WWW.WIKAPPENDIX.COM

Contents

Energy and Work
>   Derivation of New Complete Energy Equation ($E \neq mc^2$)
>   > The New "Transformic" Energy

Kinetic vs Transformic vs Potential Energy
>   Forms of Energy and a General Statement about Material Existence
>   Potential Energy and "The Constant of Integration"
>   > The "Big Bang" Connection

Gravity
>   Corporeal Gravity—Net Resultant Force from Flux Density
>   > Example—Standing on Surface of Earth
>   Dynamic Gravity—Force Caused by Angular Velocity
>   > Example—Tidal Action by Moon's Rotation about Barycenter

Forces
>   Weak Forces (Gravity)
>   Strong Nuclear Forces
>   Medium Forces—Strong Versus Medium Versus Weak
>   Force at a Distance
>   > Centrifugal Motion, $F = mv^2/r = mr\omega^2$, with changing center of r

Buoyant Force

Atomic Structure
>   A New Model of the Atom Is Required. Why?
>   > Molar Volume—Ideal Gases
>   > Temperature and Thermal Conductivity
>   > Phase Changes (solid, liquid, gas)—"Latent Heats"
>   > Ideal Gas Law—$PV = nRT$
>   > Positive and Negative Attraction/Repulsion
>   > Force at aDistance
>   > Compounds
>   > Reevaluation of Rutherford's Experiment
>   > Suggested Model—The Vacuometric Atomic Cell (VAC)
>   > > The Vacuum Core (True Vacuum)
>   > > The Speckra
>   > > Corporeal Flux—AC Phenomenon
>   > > Ethertron (compressible gaseous electron phase)
>   > > Electron (incompressible liquid electron phase)
>   > > > Boundary Layer Effect

Positive and Negative Spirogridic Opposition or
Synchronization
VAC is a Superior Model of Atomic Structure, Here is Why
Intrinsic Versus Extrinsic Energy of the Atom

3D Cymatic Formations —AC Resonance
Why the Elements Have Different Properties
Peculiarities of Water Molecule and Carbon
Periodic Table of Elements—118 Cymatic Structures
Material Structure
Physical Metallurgy
Crystallography
Eutectic Cymatic Structures
Surface Tension

Light

Simple Wave Mechanics After All
Michelson-Morley experiment revisited
All Gases Transmit Light
Ether Dilemma
Photoelectric Effect
Fudge Factor Photon
No Such Thing as a Vacuum
(Outside the VAC, Black Holes, and Dark Matter)
Invalidation of Lorentz c^2
Transparent Materials and the VAC
A true vacuum would not be transparent

Electricity

Electrical Current
Magnetism
Electrical Induction
Voltage
Electrochemical (Galvanic Series & Electromotive Potentials/
Batteries)
Induced Voltage
Electrical Conduction

Earth's Magnetic Polarity
How It Can Instantaneously Switch
     Right-Hand Rule of Electrical Induction
Brownian Motion
Cloud Chamber
More on the Ether Story
     Vacuum "Pressure" of Space calculation
     "Nothing Is Something"
     The Electron Fluid
     Compressible and/or Incompressible Medium
     The "Quantum" View/Bias

Global Warming and Global Cooling
          Equilibrium

Material Phenomena and Existence– General Statement
     Matter
     Time
     Distance
     Nothingness (True Vacuum)

ABOUT THE AUTHOR

Author Angelo Spadoni grew up in Imperial Beach, California. He is the grandson of Anacleto Spadoni of Roccafluvione, Italy, and Cleonice Mezzetti of Montefiascone, Italy. His father was Junkman John, also known as Black Jack from Onaway, Michigan, who was a self-taught particle physicist. His mother is a saintly pre-school teacher who came to the United States from Italy after World War II in 1945, who has always been willing to sacrifice her own comfort for the benefit of others. Introduced to controversial scientific principles at an early age by his father, he was also taught by his mother "soportare pasentamente le personi moleste." A registered engineer, Spadoni has a degree in metallurgical engineering from California Polytechnic, S.L.O., and specializes in electrochemistry.

WHAT IS YOUR CHAPTER 4?

The author is looking for like-minded individuals who would be interested in sharing in the quest to discover new truths in science and religion. This would include the funding of expeditions and the production of entertaining documentaries. Please contact the author at www.angelospadoni.com if you are interested in making a contribution or other arrangement.